Engineering Mechanics and Design Applications

Transdisciplinary
Engineering Fundamentals

Engineering Mechanics and Design Applications

Transdisciplinary
Engineering Fundamentals

Atila Ertas

CRC Press
Taylor & Francis Group
Boca Raton London New York

CRC Press is an imprint of the
Taylor & Francis Group, an **informa** business

CRC Press
Taylor & Francis Group
6000 Broken Sound Parkway NW, Suite 300
Boca Raton, FL 33487-2742

First issued in paperback 2018

ISBN-13: 978-1-4398-4930-9 (hbk)
ISBN-13: 978-1-138-38205-3 (pbk)

Library of Congress Cataloging-in-Publication Data

Ertas, Atila, 1944-
  Engineering mechanics and design applications : transdisciplinary engineering fundamentals / Atila Ertas.
    p. cm.
  Includes bibliographical references and index.
  ISBN 978-1-4398-4930-9 (hardback)
  1. Engineering design. 2. Mechanics, Applied. I. Title.

TA174.E7823 2012
620'.0042--dc23

2011029893

Visit the Taylor & Francis Web site at
http://www.taylorandfrancis.com

and the CRC Press Web site at
http://www.crcpress.com

To my mentors, Professor C. V. Ramamoorthy and Dr. Raymond T. Yeh, dearest friends, who inspired me to write this book.

Contents

Preface

In the last decade the number of complex problems facing engineers has highly increased and the technical knowledge and understanding in science and engineering required to address and mitigate these problems has been evolving rapidly. The world is becoming increasingly interconnected as new opportunities and highly complex problems connect the world in ways we are only beginning to understand. When we do not solve these problems correctly and in a timely manner, they rapidly become crises. Problems such as energy shortages, pollution, transportation, the environment, natural disasters, safety, health, hunger, and the global water crisis threaten the very existence of the world as we know it. Recently, fluctuating fuel prices and environmental concerns have set car manufacturers in search of new zero-polluting, fuel-efficient engines. None of these complex problems can be understood from the sole perspective of a traditional discipline. The last two decades of designing large-scale engineering systems have demonstrated that neither mono-disciplinary nor inter- or multidisciplinary approaches provide an environment that promotes the collaboration and synthesis necessary to go beyond existing disciplinary boundaries and produce truly creative and innovative solutions to large-scale, complex problems. These problems not only include the design of engineering systems with numerous components and subsystems, which interact in multiple and intricate ways, but also involve the design, redesign, and interaction of social, political, managerial, commercial, biological, medical, and other systems. Furthermore, these systems are likely to be dynamic and adaptive in nature. Obtaining the solutions to such unstructured problems requires many activities that cut across traditional disciplinary boundaries, that is, transdisciplinary research and education.

The results of transdisciplinary research and education include the following: emphasis on teamwork, bringing together investigators from diverse disciplines, and developing and sharing concepts, methodologies, processes, and tools. All of these help to create fresh, stimulating ideas that expand the range of possibilities. The transdisciplinary approach creates a desire in people to seek collaboration outside the bounds of their professional experience to make new discoveries, explore different perspectives, express and exchange ideas, and gain new insights.

This book was developed completely based on the fundamentals of engineering mechanics for undergraduate students. The mathematical level is similarly limited to that found in engineering mathematics that covers calculus and differential equations.Throughout the book, discussions of theory are followed by practical examples.

Engineering Mechanics with Design Applications: Transdisciplinary Engineering Fundamentals has been written to fulfill the need for a textbook or reference book that is appropriate for use in engineering design practice. It is intended to foster a thorough understanding of the basic condensed knowledge necessary for engineering design.

A general introduction to the concept of Prevention through Design for preventing occupational injuries, illnesses, and fatalities is provided in Chapter 1. Students should emerge from this transdisciplinary educational experience with a broad understanding of occupational safety and health needs in the design process that will prevent or minimize work-related hazards.

A condensed introduction to engineering statics is provided in Chapter 2, followed by engineering dynamics in Chapter 3. Chapter 4 deals with solid mechanics, which covers many design applications, and Chapter 5 covers failure theories and dynamic loadings. Finally, Chapter 6 integrates knowledge gained from the previous chapters for real-life design analysis and applications. In this chapter, two tragic accidents that occurred in 2010 are discussed. These two disasters killed many workers and caused property damages worth millions of dollers as well as widespread environmental damage. Therefore, the Prevention through Design (PtD) concept and some related concerns are also discussed.

This book is also geared for graduate classes covering engineering fundamentals for nonengineering students and for review by engineering students who have been out of school for several years. It will also be a valuable source for engineering students preparing for the Fundamentals of Engineering Exam and for non-mechanical engineers collaboratively working on large-scale projects. Finally, this book can be used as a textbook for non-mechanical engineering disciplines such as industrial engineering, electrical engineering, and chemical engineering.

Acknowledgment

The author would like to express his gratitude to Turgut B. Baturalp, a PhD student, who provided assistance in preparing this manuscript.

1 Prevention through Design

A Transdisciplinary Process

1.1 INTRODUCTION

The main objective of this chapter is to introduce the concept of Prevention through Design (PtD) to prevent occupational injuries, illnesses, and fatalities. Students should emerge from this transdisciplinary educational experience with a broad perspective on occupational safety and health needs in the design process that will prevent or minimize work-related hazards.

1.2 TRANSDISCIPLINARITY AND PtD

1.2.1 DISCIPLINE

Since the 1950s the integration of research methods and techniques across disciplines has been of great interest in the social and natural sciences [1]. A particular area of study is called a "discipline," provided it has cohesive tools, specific methods, and a well-developed disciplinary terminology. As disciplines inevitably develop into self-contained shells, interaction with other disciplines is minimized. However, practitioners of a discipline develop effective intra-disciplinary communication based on their disciplinary vocabulary. Many distinguished researchers and educators have contributed to the development of transdisciplinary education and research activities [2–18].

Multidisciplinary activities involve researchers from various disciplines working independently, each from their own discipline-specific perspective, to solve a common problem. Although multidisciplinary teams cross discipline boundaries, they remain limited to the framework of disciplinary research.

In interdisciplinary activities, researchers from diverse disciplines collaborate by exchanging methods, tools, concepts, and processes to find integrated solutions to common problems. Both multidisciplinary and interdisciplinary activities cross discipline boundaries, but their goal remains within the framework of disciplinary research.

1.2.2 DEFINING TRANSDISCIPLINARITY

In German-speaking countries, the term *transdisciplinarity* is used for integrative forms of research [19]. Transdisciplinary education and research programs take collaboration across discipline boundaries a step further than do multidisciplinary and interdisciplinary programs. The transdisciplinary concept is a process by which researchers from diverse disciplines work together to develop and use a shared conceptual framework to solve common problems. A distinctive characteristic of transdisciplinary research is the loosening of theoretical models and the development of a new conceptual synthesis of common terms, measures, and methods that produce new theories and models [20]. The three terms—multidisciplinary, interdisciplinary, and transdisciplinary—are often defined differently by researchers and educators.

Nicolescu [21] stated that transdisciplinarity concerns that which is simultaneously between the disciplines, across the different disciplines, and beyond all disciplines.

Klein [8] defined the above-mentioned three terms as follows: "Multidisciplinary approaches juxtapose disciplinary/professional perspectives, adding breadth and available knowledge, information, and methods. They speak as separate voices, in encyclopedic alignment. ...

Interdisciplinary approaches integrate separate disciplinary data, methods, tools, concepts, and theories in order to create a holistic view or common understanding of complex issues, questions, or problem. ... Theories of interdisciplinary premised on unity of knowledge differ from a complex, dynamic web or system of relations.

Transdisciplinary approaches are comprehensive frameworks that transcend the narrow scope of disciplinary world views through an overarching synthesis, such as general systems, policy sciences, feminism, ecology, and sociobiology. ... All three terms evolved from the first OECD international conference on the problems of teaching and research in universities held in France in 1970."

Hadorn et al. [22] stated, "Transdisciplinary research is research that includes cooperation within the scientific community and a debate between research and the society at large. Transdisciplinary research therefore transgresses boundaries between scientific disciplines and between science and other societal fields and includes deliberation about facts, practices and values."

Peterson and Martin [23] stated that interdisciplinary research has not produced a combination or synthesis that would go beyond disciplinary boundaries to produce innovative solutions to policy questions. However, they proposed that transdisciplinary approaches call for a synthesis of research at the stages of conceptualization, design, analysis, and interpretation by using integrated team approaches.

Stokols et al. [24] defined transdisciplinary science as collaboration among scholars representing two or more disciplines in which the collaborative products reflect an integration of conceptual and/or methodological perspectives drawn from two or more fields.

Petts et al. [25] stated, "One of the broadly agreed characteristics of transdisciplinary research is that it is performed with the explicit intent to solve problems that are complex and multidimensional, particularly problems (such as those related to sustainability) that involve an interface of human and natural systems."

During the past decade, other approaches of transdisciplinarity have been developed and described by several researchers and educators. Common phrases in the definitions above include collaboration, shared knowledge, unity of knowledge, distributed knowledge, common knowledge, integration of knowledge, new knowledge generation, integrated disciplines, beyond discipline, complex problems, and societal fields. Although a precise definition of transdisciplinarity is debatable, by reviewing the above approaches, definitions, and common phrases, the following definitions can be put forward [18]:

Transdisciplinarity is the development of new knowledge, concepts, tools, and technologies shared by researchers from different disciplines (social science, natural science, humanities, and engineering). It is a collaborative process of organized knowledge generation and integration by crossing disciplinary boundaries to design and implement solutions to unstructured problems.

Transdisciplinary knowledge is a shared, common, collective knowledge derived from diverse disciplinary knowledge cultures (engineering, natural science, social science, and humanities).

The transdisciplinary research process is the collaboration among scholars from diverse disciplines to develop and use integrated conceptual frameworks, tools, techniques, and methodologies to solve common unstructured research problems. Transdisciplinary research creates new paradigms and provides pathways to new frontiers.

The fundamental characteristics of transdisciplinary research are

- Use of shared concepts, frameworks, tools, methodologies, and technologies to solve common unstructured research problems
- Elimination of disciplinary boundaries for strong collaboration
- Redefinition of the boundaries of natural science, social science, humanities, and engineering by building bridges between them
- Development of shared conceptual frameworks, new knowledge, tools, methodologies, and technologies

1.2.3 MULTIDISCIPLINARY, INTERDISCIPLINARY, AND TRANSDISCIPLINARY CASE STUDIES

Wind power promises a clean and inexpensive source of electricity. It promises to reduce our dependence on imported fossil fuels and to reduce the output of greenhouse gases. Many countries are, therefore, promoting the construction of vast "wind farms" and encouraging private companies by providing generous subsidies. The goal of the U.S. Department of Energy (DOE) in 2010 was to produce 5% of the country's electricity by wind turbine farms. The history of wind power shows a general evolution from the use of simple, lightweight devices to heavy, material-intensive drag devices and finally to the increased use of lightweight, material-efficient aerodynamic lift devices in the modern era.

During the winter of 1887–1888, Charles F. Brush built the first automated wind turbine for generating electricity. It was the world's largest wind turbine with a rotor diameter of 17 m (50 ft) and 144 rotor blades made of cedar wood. The turbine ran for 20 years; the batteries, Which were charged in the cellar of Brush's mansion, lasted even longer [26].

Wind has been an important source of energy in the United States for some time. Over 8 million mechanical windmills have been installed in the country since the 1860s. It is interesting to note that some of these units have been in operation for more than a hundred years [27].

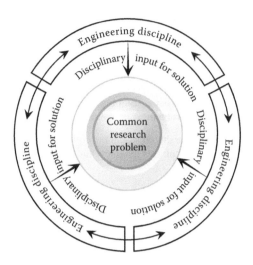

FIGURE 1.1   Multidisciplinary research process.

A wind turbine system consists of subsystems to catch the energy of the wind by converting mechanical rotation into electrical power Systems in place also start, stop, and control the turbine. To design today's large wind turbine structures, many researchers from different disciplines need to work together: mechanical engineers working on gear design, civil engineers designing the structure, material engineers selecting the most suitable material for application, electric engineers working on power transmission and control system design, and wind engineers designing the rotor blade. Each subcomponent is designed independently and the whole system is put together. But does this process provide an optimum design? The answer seems to be an obvious no! As shown in Figure 1.1, the common research problem is to design a wind turbine. While a simple methodology would be to create a collaborative research team using the multidisciplinary research process, multidisciplinary teams tend to remain limited to the framework of disciplinary research. Better collaboration and organization is necessary for such complex system designs.

If the research approach is interdisciplinary, as shown in Figure 1.2, researchers from different engineering disciplines communicate with each other to optimize their subcomponent design keeping in mind requirements of the whole system design. Once the compatibility and reliability of the subcomponents are ensured, the system can be assembled. This approach seems to provide an integrated solution to a common problem. However, the question remains as to whether this process results in an optimum design for the use of wind power? Again, the answer perhaps is no.

Although wind power promises a clean and inexpensive source of electricity, wind turbines can raise environmental and community concerns. For example:

- Noise and vibrations from wind turbines can cause sleep disruptions and other health problems among people who live nearby.
- Wind turbines can be visually intrusive for residents living near them.

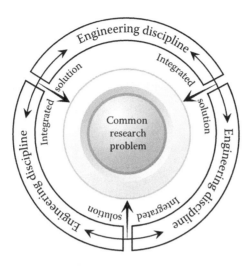

FIGURE 1.2 Interdisciplinary research process.

- Wind turbines can disturb wildlife habitats and cause injury or death to birds.
- Air turbulence from wind farms could adversely affect the growth of crops in the surrounding countryside.
- Huge wind turbines, taller than a 60-story building, with blades more than 300 ft long may cause disturbance to the community.

In the late 1980s, the California Energy Commission reported that 1300 birds were killed by wind turbines, including over 100 golden eagles at Altamont Pass, California. Environmental issues related to wind turbines include impact on wildlife, habitat, wetlands, dunes, water resources, soil erosion, and sedimentation. Other areas of concern are interference with TV and microwave reception, depreciating property values, increased traffic, road damage, cattle being frightened from rotating shadows cascading from the turbine blades, rotating shadows in nearby homes, stray voltage, increased lightning strikes, and so on. Currently, all of these issues are being raised in states where wind farms have been introduced.

These concerns confirm that while an interdisciplinary approach to wind turbine design can result in a more efficient turbine it does not provide a comprehensive solution for the effective use of wind energy.

As shown in Figure 1.3, the transdisciplinary process not only involves crossing boundaries within different fields of engineering but also requires crossing boundaries between different disciplines (engineering, social science, natural science, and humanities). Social sciences and the humanities bring an abundance of knowledge on cultural, economic, and social growth and advancement as well as on social systems. Therefore, they provide perspectives for decisions being made with regard to current problems and challenges. The humanities play an important role by putting to beneficial use new findings in the engineering and natural sciences. For example, natural scientists work together with researchers in the humanities to discover archaeological objects and determine their age.

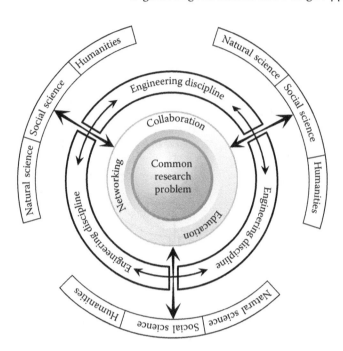

FIGURE 1.3 Transdisciplinary research process.

For the development of wind turbine farms, environmental science researchers should make an assessment of the site based on a comprehensive consultation with local community and environmental bodies. Engineers should work with researchers from the social sciences, natural sciences, and the humanities to understand the impact on the environment and the neighboring communities to guide their design.

Through the transdisciplinary approach, researchers can plan early and consult with the affected communities frequently. This will allow them to identify and address the most serious issues before substantial investments are made. In other words, design engineers should make reasonable efforts to "design out" or minimize hazards and risks early in the design process.

Further, researchers from diverse disciplines should work with the required utility agencies, government agencies, environmental organizations, and developers to ensure that any complex problems are under control.

Continuous education and encouragement develop a spirit of collaboration among researchers to help solve these complex problems. Through educational activities that focus on research team management, problem solving, establishing research goals, optimizing the use of resources, and supporting each other, members of the research team learn to work together more effectively. In other words, team members mentor and support each other. For transdisciplinary teams to be effective, they must meet on a regular basis. Members of transdisciplinary teams have a large information network and extended contacts who are capable of collaborating on a project from the beginning to the final implementation stage.

A transdisciplinary research community is a network of researchers from diverse disciplines.

As shown in Figure 1.3, collaboration, networking, and education on a global scale are the keys to solving the complex problems facing humankind in this century. The successful development of a network of global collaboration centers and institutes will provide common sharing of knowledge that will benefit society by significantly enhancing our ability to the solve unstructured problems the world is facing today.

1.2.4 WHY IS PREVENTION THROUGH DESIGN A TRANSDISCIPLINARY PROCESS?

Paul A. Schulte, director of the Education and Information Division of the National Institute for Occupational Safety and Health (NIOSH), stated that the "Prevention through Design (PtD) process is a collaborative initiative that lies on the principle that the best way to prevent occupational injuries, illnesses, and fatalities is to anticipate and 'design out' or minimize hazards and risks when new equipment, processes, and business practices are developed" [28]. He also emphasized that the PtD process requires cross-disciplinary activities.

Fisher [29] reported that "implementing PtD will require the challenging transformative concept. Transformative changes are broad[er] and can lead to new forms and practices that guide us to safer and more productive environments. PtD, if viewed and practiced with broad vision, should further transformative changes that promote patient, worker, and environmental safety." A number of similarities exist between transformative and transdisciplinary concepts.

Schulte et al. [30] stated, "An important element that should be included in the initiative is the need for global cooperation or harmonization. Due to the global influence on economies, workplaces, designs, and occupational safety and health, any major initiative, such as PtD, needs to have global input and support." Since PtD directly and indirectly involves global issues, strong international collaboration and partnerships need to be established among stakeholders to have worldwide input and support for PtD. This important observation reveals that PtD is a transnational activity.

The American workforce has undergone a significant change because of immigration. Immigrants with job opportunities in the United States usually have lower educational skills, are poorer, and earn less than the native population. In this situation, the difficulties of developing culturally integrated approaches to workplace safety and health should not be underestimated. As the world becomes increasingly multicultural, PtD processes should consider synthesized transcultural theories, models, and research to facilitate culturally harmonious prevention and control of occupational injuries, illnesses, and fatalities.

The above discussions reveal that PtD is a concept shared by diverse disciplines, including agriculture, forestry, and fishing; healthcare and social assistance; mining; services; construction; manufacturing; transportation, warehousing, and utilities; and wholesale and retail trade. A common research problem addressed by PtD in these diverse disciplines is prevention and control of occupational injuries, illnesses, and fatalities.

Thus, PtD is a transdisciplinary process that involves many transnational and transcultural issues.

1.2.5 P**T**D C**ONSIDERATIONS FOR THE** D**ESIGN** P**ROCESS**

1.2.5.1 Generic Design Process

The typical steps in the engineering design process shown in Figure 1.4 are considered to be applicable to most design efforts, but the reader should recognize that individual projects often require variations, including the elimination of some steps [31].

1.2.5.1.1 Recognition of Need and Requirements

The design process begins with an identified need that can be satisfied by defined customer requirements, design requirements, and functional requirements. During this phase, the design team works closely with the customer to determine the requirements for the product, identifying the functionality, performance levels, and other characteristics that the product must satisfy. The requirements ascertained in this phase serve as a foundation for the remaining phases of the design process. It is important to note that the establishment of valid design requirements is re-examined during the preliminary design phase.

1.2.5.1.2 Conceptual Design Phase

After the problem has been adequately defined during the concept development phase, viable solutions need to be identified from which the optimum approach can be selected. For small-scale projects, an assessment of the feasibility of the selected

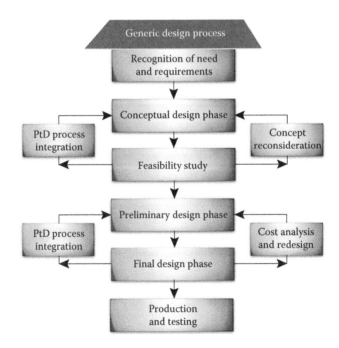

FIGURE 1.4 Generic design process. (Adapted from Ertas, A. and Jones, J. 1996. *The Engineering Design Process*, 2nd edition, John Wiley & Sons, Inc., New York.)

concept(s) is often made as part of this phase; however, for larger projects, this assessment is usually a major element of the overall program and sometimes it can take several years to complete. Assessing the feasibility of the concept(s) ensures that the project proceeds into the design phase with a concept that is achievable, both technically and within cost constraints, and that new technology is required only in areas that have been thoroughly examined and agreed to. It is important to have well-experienced research team members judge/assess the feasibility of the design process. Team members in charge of this assessment phase should be directly responsible for the overall performance and functionality of the product, process, or facility—people who have a *work ownership* mentality.

1.2.5.1.3 Preliminary Design Phase

The preliminary design phase may also be known as architectural design. This phase bridges the gap between the conceptual design phase and the detailed design phase. The design concept is further defined during the preliminary design phase, and if more than one concept is involved, an assessment leading to the selection of the best overall solution must be performed. System-level and, to the extent possible, component-level design requirements should be established during this phase of the process. The overall system configuration is defined during the preliminary design phase, and a schematic, diagram, layout, drawing, or other engineering documentation (depending on the project) should be developed to provide early project configuration control. This documentation will assist in ensuring interdisciplinary or transdisciplinary team integration and coordination during the detailed design phase. The preparation of system testing and operational and maintenance procedures at an early stage in the design also often helps. The process of thinking these procedures through helps to quantify the various design parameters and thus provide a valid basis for component design.

1.2.5.1.4 Detailed Design Phase

The goal of the detailed design phase is to develop a system of design drawings and specifications that provides a detailed specification for each component, thoroughly describing interfaces and functions of each component so that it can be manufactured. At this phase, all the designers and researchers from the diverse disciplines are actively involved in the synthesis/analysis process, distinguishing component parts of the system design concept, evaluating components to validate previously established requirements, specifying those design requirements left undefined, and assessing the effect of the component requirements on the overall system requirements. The detailed design phase serves as the basis for the production phase.

1.2.5.1.5 Production and Testing Phase

During this phase of the project, using the specifications created in the previous phases, the actual product is developed and manufactured. The final product is then tested to ensure that it meets the requirements defined in the requirements phase. As shown in Figure 1.5, PtD should be an important consideration throughout the design process.

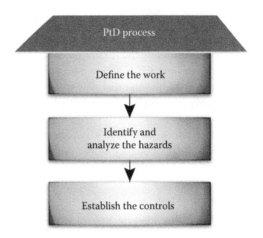

FIGURE 1.5 PtD process.

1.2.5.2 PtD Process

The goal of this section is the integration of PtD considerations into design activities during the conceptual, preliminary, and final design stages. Figure 1.5 shows a general process for PtD: defining the work related to product design, identifying and evaluating potential safety hazards and injuries involved with the product, and finally controlling hazards that cannot be eliminated. This activity should be implemented throughout the entire design process as shown in Figure 1.4.

PtD must be fully integrated in the early design process in the project. At the beginning of concept development, a hazard analysis of alternatives to be considered and worker safety and health requirements for the design must be established. The main objective of PtD at the conceptual design phase is to evaluate alternative design concepts, to plan to protect workers from safety and health hazards, and to provide a conservative safety design basis for a chosen concept to be taken forward into preliminary design. The conceptual design phase offers a key prospect for the safety and health hazard analysis to influence product design.

PtD efforts during the preliminary design phase should be incremental, that is, assessing fundamental concerns at each level, instead of merely re-examining the complete conceptual design. Hazard analysis should progress from a facility-level analysis to a system-level analysis as design detail becomes available. During hazard analysis, the selection of controls, safety considerations, and classifications developed during the conceptual design phase must be revisited to make sure they are still appropriate. Decisions made during the preliminary design phase provide the basis for the approach to detailed design and production.

During the detailed design phase, a final set of hazard controls should be developed based on hazard analysis. More detailed information on this subject can be found elsewhere [32].

The National Safety Council has recommended basic guidelines for designers to ensure acceptable safety and health precautions for products and processes. The guidelines given below are broad, and as many as possible should be considered during product design and use [31]:

- Eliminate hazards by changing the design, the materials used, or the maintenance procedures
- Control hazards by capturing, enclosing, or guarding at the source of the hazard
- Train personnel to be cognizant of hazards and to follow safe procedures to avoid them
- Provide instructions and warnings in documentation and post them in appropriate locations
- Anticipate credible abuse and misuse and take appropriate action to minimize the consequences
- Provide appropriate personal protective equipment (PPE) and establish procedures to ensure that it is used as required

Engineers must be able to identify hazards related with their product designs and to quantify the relative severity and probability of occurrence. Safety hazards normally result in accidents that occur over a relatively short period of time and for which severe effects are readily apparent. The effects of health hazards, in contrast, may not be obvious for some time, often months or years, but the results can be just as damaging [33]. A number of techniques have been proposed to help recognize, quantify, and reduce hazards. Haddon's 10 rules given below comprise one of the more commonly recognized strategies [34]:

1. Prevent the creation of the hazard (e.g., prevent the production of hazardous and nonbiodegradable chemicals).
2. Reduce the magnitude of the hazard (e.g., reduce the amount of lead in gasoline).
3. Eliminate hazards that already exist (e.g., ban the use of chlorofluorocarbons).
4. Change the rate of distribution of a hazard (e.g., control the rate of venting a hazardous propellant).
5. Separate the hazard from that which is being protected (e.g., store flammable materials at isolated locations).
6. Separate the hazard from that which is being protected by imposing a barrier (e.g., separate fuel and oxidizer storage areas by using beams or other barriers).
7. Modify basic qualities of the hazard (e.g., use breakaway roadside poles).
8. Make the item to be protected more resistant to damage from the hazard (e.g., use fabric materials in aircraft that do not create toxic fumes when combusted).
9. Counter the damage already done by the hazard (e.g., move people out of a contaminated area).
10. Stabilize, repair, and rehabilitate the object of the damage (e.g., rebuild after a fire).

Even though the above guidelines and rules do not include every possible safety consideration in a design project, they do provide a checklist against which the design can be evaluated and modified as needed. The designer must continuously evaluate the design for safety, considering not only the product design but also the workers involved in fabricating the product, maintaining and repairing the product or system, as well as the end user or purchaser.

Developing the manufacturing processes as well as the maintenance and operating procedures early during the design process will help reveal safety problems at a time when corrective action can be taken at minimum cost.

In the engineering design, development, and fabrication processes, there are some seldom-used techniques available that help to ensure avoiding unwelcome and sometimes very costly surprises during final assembly and testing. One of these techniques involves the preparation of assembly, testing, and operational procedures concurrently with the design process. This is difficult because the design details are just being formulated and are highly subject to change; nevertheless, the value of looking ahead to identify assembly methods and potential operational problems can prove to be vital. This did occur during a large (and costly) missile system development program in the late 1950s. There was a misunderstanding between two of the major design groups regarding the system design requirements. The mechanical group had made provisions to operate the system in any one of three modes, but the electrical group's design allowed for only one mode of operation. This did not become apparent until very late in the program when two engineers from the mechanical group were assigned to check the overall system and test procedures. They found that the system could only be operated in a fully automatic mode; operating the system in a semiautomatic or manual mode was impossible. This was a very embarrassing turn of events for the contractor, and it undoubtedly did considerable damage to their reputation with the customer. This could have been circumvented had the overall system been checked and procedures tested early in the design process, and the design process been continuously updated.

Another technique that costs little in terms of effort or expense but offers great advantages in certain situations during design and development programs is the use of simple models. Inexpensive models can be developed to answer problems associated with assembly of complicated parts, to evaluate the feasibility of certain operations, and to provide a visual conception of size related to function. A good example of this occurred during a large missile emplacement program in the 1960s. A problem arose when a contractor who was installing propellant piping stated that he could not get a large section (approximately 20 ft long, with a complex configuration) of the piping into the missile silo. The propellant piping was loaded into the silo by a crane through a passageway on the side of the silo. The propellant piping was fabricated in California, then cleaned and sealed under a low blanket pressure, and shipped to various sites around the United States. Thus, it would have been very costly to redesign and fabricate replacement pipe sections. The assessment of the contractor was subsequently challenged by the customer after conferring with his propellant system consultant. Fortunately, there was a civil engineer in the consultant group who recommended that a model be constructed to determine whether it was possible to rig the section of piping so that it could pass through the passageway. With considerable doubt about the possibility of answering a question involving such minute measurements using an inexpensive model to replicate such a large structure, the propellant consultant worked with the civil engineer to construct a model of the silo and passageway from cardboard. The piping section was modeled using a wire pipe cleaner. Sure enough, the model verified that the piping section could not be passed into the silo through the passageway when rigged as the contractor had proposed. However, when rotated 180° and rigged so that the opposite end of the piping section entered the silo first, the model piping section would just clear the passageway and slip into the silo [31].

1.3   PREVENTION THROUGH DESIGN

PtD is a process or concept used to prevent and control occupational injuries, illnesses, and fatalities or to reduce workplace safety risks and lessen workers' reliance on PPE [35]. In other words, PtD is a process of integration of hazard analysis and risk assessment methods early in the design and engineering stages, leading to subsequent actions necessary to prevent risks of injury or damage.

Several national organizations have partnered with NIOSH in promoting this concept of recognizing the hazards of each industry and designing more effective prevention measures. These national organizations include the American Industrial Hygiene Association, American Society of Safety Engineers, Center to Protect Workers' Rights, Kaiser Permanente, Liberty Mutual, National Safety Council, Occupational Safety and Health Administration, ORC Worldwide, and Regenstrief Center for Healthcare Engineering.

Work-related injuries are real, devastating, and common. Recent studies indicated that each year in the United States, 55,000 people die from work-related injuries and diseases, 294,000 take sick, and 3.8 million are injured. Annual direct and indirect costs have been estimated to range from $128 billion to $155 billion. Recent studies in Australia indicate that design is a significant contributor in 37% of work-related fatalities; hence, the successful implementation of PtD concepts can have substantial impact on worker health and safety [36].

The concept of PtD can be defined as:

> Addressing occupational safety and health needs in the design process to prevent or minimize the work-related hazards and risks associated with the construction, manufacture, use, maintenance, and disposal of facilities, materials, and equipment [35].

Today, many business leaders are recognizing PtD as a cost-effective means to enhance occupational safety and health, and are therefore openly supporting the PtD process and developing effective management practices to comply with PtD.

Besides the United States, many other countries are actively promoting PtD concepts as well. For example, in 1994, the United Kingdom needed construction companies, project owners, and architects to address safety and health issues during the design phase of projects, and companies there have responded with positive changes in management practices to comply with the regulations. The Australian National OHS Strategy 2002–2012 put "eliminating hazards at the design stage" as one of five national priorities. Consequently, the Australian Safety and Compensation Council developed the Safe Design National Strategy and Action Plans for Australia, encompassing a wide range of design areas, including buildings and structures, work environments, materials, and plant (machinery and equipment) [35].

The goal of PtD is to reduce the risk of occupational injury and illness by integrating decisions affecting safety and health at all stages of the design process. To move toward fulfillment of this mission, John Howard, MD, 2002–2008 Director of NIOSH, said, "One important area of emphasis will be to examine ways to create a demand for graduates of business, architecture and engineering schools to have basic knowledge in occupational health and safety principles and concepts" [37].

Although PtD initiative focuses on design, one should realize the importance of other factors, such as behavior, management, leadership, and PPE. These factors may interact directly with designs that address occupational safety and health [30].

1.3.1 PtD Program Mission

The mission of the PtD national initiative is to prevent or reduce occupational injuries, illnesses, and fatalities through the inclusion of prevention considerations in all designs that impact workers. The mission can be achieved by

- Eliminating hazards and controlling risks to workers to an acceptable level "at the source" or as early as possible in the life cycle of items or workplaces
- Including design, redesign, and retrofit of new and existing work premises, structures, tools, facilities, equipment, machinery, products, substances, work processes, and the organization of work in PtD program mission
- Enhancing the work environment through the inclusion of prevention methods in all designs that impact workers and others on the premises

The program strives to fulfill its mission through the following principles:

- High-Quality Research: NIOSH will continually strive for high-quality research and prevention activities that will lead to reduction in occupational injuries and illnesses among workers in the PtD cross-sector.
- Practical Solutions: The NIOSH program for the PtD cross-sector is committed to the development of practical solutions to the complex problems that cause occupational diseases, injuries, and fatalities among workers in this sector. One source of practical recommendations is the NIOSH Health Hazard Evaluations (HHE) program. NIOSH conducts HHEs at individual worksites to find out whether employees face health hazards caused by exposures or conditions in the workplace.
- Partnerships: Collaborative efforts in partnership with labor, industry, government, and other stakeholders are usually the best means of achieving successful outcomes. Developing these partnerships is a cornerstone of the NIOSH program for the PtD cross-sector.
- Research to Practice: Research only realizes its true value when put into practice. Every research project within the NIOSH program for the PtD cross-sector formulates a strategy to promote the transfer and translation of research findings into prevention practices and products that will be adopted in the workplace.

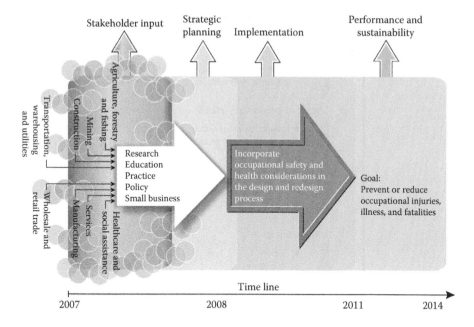

FIGURE 1.6 Approach toward the development and implementation of the PtD national initiative process. (Redrawn from Schulte, P. A. and Heidel, D. S. 2009. *Prevention through Design in Motion*, June 17.)

1.3.2 PtD Process

Figure 1.6 shows that the PtD process prevents and controls occupational injuries, illnesses, and fatalities [39]. As shown in the figure, stakeholders whose input is needed in the PtD process are agriculture, forestry, and fishing; healthcare and social assistance; mining; services; construction; manufacturing; transportation, warehousing, and utilities; and wholesale and retail trade. Figure 1.7 shows a flow chart for the PtD process [30].

1.3.3 Stakeholder Input

The plan, developed from stakeholder input, focuses on eliminating hazards and minimizing risks in all designs affecting workers (more information on the plan developed from stakeholder input is available on the NIOSH website). Work on the plan was initiated at the first PtD workshop in Washington, DC on July 9–11, 2007. In all, 250 stakeholders/workshop participants from diverse industry sectors, including representatives from labor, industry, academia, and government discussed the most compelling issues to be included in this plan. As Figure 1.6 depicts, NIOSH is now in the implementation phase of the PtD plan.

Figure 1.8 shows examples of PtD in each sector [39]. For example, Figure 1.8d shows how to prevent injury through effective design.

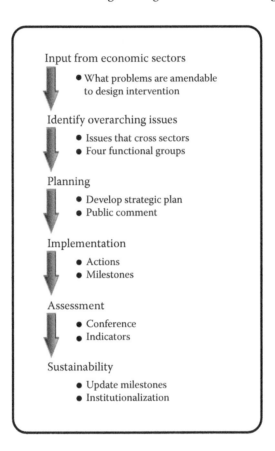

FIGURE 1.7 PtD process flow chart. (Adapted from Schulte, P. A. et al. 2008. *Journal of Safety Research*, 39, 115–121.)

Hazards associated with lifting (transferring and repositioning patients) cause significant risk to healthcare workers in the United States. Studies show that, in 2003 alone, caregivers suffered 211,000 occupational injuries. As the population ages and the demand for skilled care services increase, the occurrences of musculoskeletal injuries to the back, shoulder, and upper extremities of caregivers also increase [39]. Healthcare workers experience musculoskeletal disorders caused by heavy manual lifting at a rate exceeding that of workers in construction, mining, and manufacturing [40], and, as shown in Figure 1.8d, the risk of back injuries can be reduced/prevented by properly designed patient-lifting devices and patient safety and comfort can be improved.

Agriculture ranks fourth in the United States for work-related fatalities [41]. In general, fatalities associated with agricultural machinery involve farm tractors and rollover incidents. As shown in Figure 1.8a, death and injuries can be prevented with built-in rollover protection.

"Sharps" include needles, syringes, razor blades, slides, scalpels, pipettes, broken plastic, glassware, and other objects capable of cutting or piercing the skin [42].

Agriculture, forestry and fishing   Healthcare and social assistance

    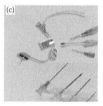 

Preventing death and injuries   Preventing amputations by   Preventing infectious diseases   Preventing back injuries with
with built in roll-over   using an auto e-stop device   in healthcare workers through   patient lifting devices
protection   on fishing boat   safer sharps design

Construction   Mining

Preventing falls and fatalities Preventing exposures to asphalt   Preventing back injuries with   Preventing hearing loss with
with tie-in scaffolding   fumes with emmision controls   mechanical lifting devices   a coated chain conveyor

Transportation, warehousing and utilities   Wholesale and retail trade   Services

Preventing death and injuries   Preventing aircraft incidents   Preventing injuries from falling   Preventing electrocutions and
to transportation workers   by reducing work hours   merchandise using designs   falls with 3-dimensional roller
through vehicle design   in rack-guard netting   arms

FIGURE 1.8   Examples of PtD in each sector. (Adapted from Schulte, P. A. and Heidel, D. S. 2009. *Prevention through Design in Motion*, June 17.)

Sharps injuries and infectious diseases can occur in all aspects of clinical and operating room duties. Unfortunately, a "no needle, no risk" situation is not possible in all healthcare settings. However, infectious diseases in healthcare workers can be prevented through safer sharps design (Figure 1.8c).

Falls from scaffoldings are among the most frequent cause of very serious injuries, including spinal cord injury, brain injury, and even death. As shown in Figure 1.8e, by designing scaffolds with proper guard rails and toe boards, falls and fatalities can be prevented.

As shown in Figure 1.8f, over half a million workers are exposed to fumes from asphalt, a petroleum product used extensively in road paving, roofing, siding, and concrete work. Health hazards that can affect workers from exposure to asphalt fumes include headache, skin rash, sensitization, fatigue, reduced appetite, throat and eye irritation, coughing, and skin cancer. With emission controls, exposure to asphalt fumes can be prevented.

Exposure to loud noise is one of the most common causes of hearing loss. Exposure to noise levels above 85 decibels can damage hearing. The noise from power lawnmowers, tractors, and hand drills is in the range of 90–98 decibels. If workers are regularly exposed, for a minute or longer, to bulldozers, chain saws, ambulance sirens, or jet engine takeoffs, they are at risk of hearing loss. Recent studies show that 30 million people are at risk in the workplace, in recreational settings, and at home. Hearing loss is the most common work-related health hazard. For example, as shown in Figure 1.8h, applying coatings on a noisy chain conveyor belt can prevent hearing loss.

1.3.4 STRATEGIC GOAL AREAS

As shown in Figure 1.9, the PtD national initiative is organized around four functional areas: research, education, practice, and policy. Small business was added as an additional focus area to address the challenges of applying PtD methods to small business processes and environments. Detailed discussions of these functional areas for eight sectors are given in Refs. [43–50].

The expected result of PtD is to prevent or reduce occupational injuries, illnesses, and fatalities through these functional areas. Each of these functional areas is supported by a strategic goal. A summary of the strategic goals for each of these areas is given in Figure 1.9 [39].

Research: PtD research is central to the NIOSH national PtD initiative. Research is required to support PtD efforts in all the other three functional areas (education, practice, and policy) and within all eight stakeholder sectors. Research will provide

PtD Strategic Goals

Research	Research will establish the value of adopted PtD interventions, address existing design-related challenges, and suggest areas for future research.
Education	Designers, engineers, health and safety professionals, business leaders, and workers understand PtD principles and apply their knowledge and skills to the design of facilities, processes, equipment, tools, and organization of work.
Practice	Stakeholders access, share, and apply successful PtD practices.
Policy	Business leaders, labor, academics, government entities, and standard developing and setting organizations endorse a culture that includes PtD principles in all designs affecting workers.
Small business	Small businesses have access to PtD resources that are designed for or adapted to the small business environment.

FIGURE 1.9 PtD strategic goals for functional areas. (Adapted from Schulte, P. A. and Heidel, D. S. 2009. *Prevention through Design in Motion*, June 17.)

the opportunity to explore and gain further understanding of the PtD concept and the evidence to support a national PtD initiative [51]. Research should take into consideration not only worker safety and health but also other prospects such as cost, quality, and sustainability.

Education: Education is a major factor required to make the PtD initiative successful. Development of PtD knowledge and skills can occur through enhanced design and engineering curricula as well as through improved professional accreditation programs that value PtD issues and include them in their performance evaluation and competency assessments. Educational requirements can be different within each sector. Therefore, an education strategy should be developed with an overall approach and set of resources, which can be customized to each sector [39,52].

The main objectives of the PtD education functional area are to [52]:

- Classify every educational action as developing awareness or capability
- Develop educational resource materials that are available to all
- Tailor approach and resources to each of the constituents in a sector
- Incorporate elements for companies of all sizes
- Conduct assessment and continuously improvement efforts
- Identify drivers of educational change and work with them

Although stakeholders recognize the need to incorporate the PtD concept in engineering course content, currently, most of the curricula of the various engineering disciplines do not include the tools and techniques needed for utilizing PtD concepts. The NIOSH has a project to diffuse PtD principles into engineering textbooks as they are being written or revised for new editions.

Practice: There are numerous examples of PtD that exemplify current business practices. These examples implementing PtD concepts should be shared with/marketed to other businesses to enhance their performance and improve their outcome. Products that are developed keeping in mind PtD concepts can become good advertising tools. Practice should also include the value of workers' health and safety in design decisions, exploring relationships with green and sustainable designs [39,53].

Policy: The policy functional area of the PtD concept includes internal and external initiatives that are planned to integrate PtD into business and governmental organizations. Research areas of PtD determine what works best, practice develops tools and implementation plans, and education teaches those who can implement PtD. PtD functional areas are not separated by clear and distinct boundaries. Instead there is significant overlap and interdependence among them [54]. "Policy focuses on creating demand for safe designs for workers and incorporating these safety and health considerations into guidance, regulations, recommendations, operating procedures, and standards" [39].

Small Business: Many large retailers have successful safety policies and programs in place. Simplified versions of such policies should be shared with/marketed to other businesses, and adapted as necessary to meet the unique needs of small businesses.

1.3.5   Business Value of PtD

Annually, 5800 people die from work-related injuries and diseases, 228,000 become ill, and 3.9 million are injured in the United States [55]. Direct and indirect costs of work-related injuries, illnesses, and fatalities have been estimated to range from $128 billion to $155 billion each year [30]. Recent studies reveal that the successful implementation of PtD concepts can greatly improve worker health and safety and reduce work-related fatalities, resulting in lower workers' compensation expenses.

In the "business case" for Industrial Hygiene (IH) study, a strategy that enables IH professionals to qualitatively and quantitatively analyze the business value of IH activities and programs was developed and tested [23,56]. The study findings demonstrated that considerable business cost savings are associated with hazard elimination and the application of engineering controls to minimize risks [39]. Occupational hygiene programs or hazard control measures provide a return on investment in the form of financial or other benefits.

Figure 1.10 illustrates PtD using a hierarchy of controls. As shown in this figure, if one selects hazard control measures higher in the hierarchy of controls, the business value increases. PtD is a concept or process that starts with identifying the hazard(s) and then eliminates the hazard(s), reducing the risk to an acceptable level by applying the hierarchy of controls shown in Figure 1.10. Elimination of potential sources of hazard(s) ranks highest in the hierarchy of controls, followed by substitution, engineering controls, and warnings.

Administrative controls are required work practices and policies that prevent hazard(s) and reduce risk. These rank low in the hierarchy of controls because their effectiveness depends on consistent implementation by management and personnel. PPE is the last line of defense and ranks lowest in the hierarchy of controls. The time to control hazards should be considered during all stages of the design process for any project.

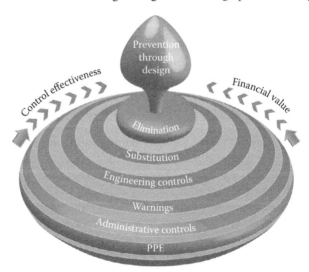

FIGURE 1.10   PtD using hierarchy of controls. (Redrawn from Schulte, P. A. and Heidel, D. S. 2009. *Prevention through Design in Motion*, June 17.)

1.4 CASE STUDIES

The following case studies demonstrate considerable savings in business costs associated with hazard elimination and the application of engineering controls to minimize risks [57].

1.4.1 CASE STUDY 1: CHEMICAL CONTAINMENT

1.4.1.1 Description of Operation

The operation where the intervention occurred is a step in the manufacturing of active pharmaceutical ingredients that were subsequently formulated in various drug products. The current operation was an open process involving the repacking of resin columns using an acetonitrile (ACN) slurry.

1.4.1.2 Hazard Identification

The current operation involved the addition of ACN into an open manway of a process tank. During the operation two operators were exposed to levels of ACN ranging from 60 to 100 parts per million (ppm). Operators were required to wear powered air-purifying respiratory protective equipment (RPE) to protect against airborne ACN exposures during the solvent charging process.

1.4.1.3 Hazard Intervention

To reduce exposure, an engineering control consisting of purchasing and installing a high-containment valve was implemented. By using the high-containment valve for charging the tank, airborne exposures of ACN were virtually eliminated.

1.4.1.4 Impacts of the Intervention

The installation of engineering controls reduced the airborne levels of ACN from the 60 to 100 ppm range to ≤1 ppm. The resultant exposure level eliminated the requirement for operators to wear RPE. As a result, there was a cost savings linked to the elimination of the RPE as well as to the associated time required to properly don/doff the RPE. Prior to the intervention, the operation step required three operators, which was subsequently reduced to two operators after the implementation of the containment project, thus significantly reducing overall labor costs associated with the operation.

Although no quality deviations had been previously associated with this manufacturing step, the containment and enclosure of the open process were also recognized as improvements in quality control. In addition, containing the process also eliminated foaming issues, sometimes noted during the operation of the process; however, the benefit of the reduction has yet to be fully evaluated.

The change in process also reduced by one third the amount of ACN lost to the environment during the operation, thus allowing a small material savings and a corresponding lowering of volatile organic compound (VOC) air emissions. The enclosed process would require additional leak detection and repair (LDAR) monitoring points to be added to the environmental monitoring schedule, but the incremental costs were minimal. Another benefit of the project was the elimination of the need to dispose used RPE as hazardous waste: one drum of hazardous waste per month and the associated disposal costs were eliminated.

1.4.1.5 Financial Metrics

The financial metrics associated with the intervention indicated that the project yielded a 5-year net present value (NPV) of $23,629 with an internal rate of return of 14%. The project had a discounted payback period (DPP) of 3.8 years. Therefore, in addition to the benefits of lower employee ACN exposures, improved air quality, reduced air emissions, and reduced hazardous waste, the project also yielded a competitive rate of return from the organization's investment. The project also resulted in some improvement in employee morale by eliminating the need to wear RPE.

1.4.1.6 Lessons Learned

The implementation of engineering controls resulted in a process change that reduced labor and material costs, improved product quality, reduced air emissions, and reduced the volume of hazardous waste generated and its associated disposal cost and liability.

1.4.2   CASE STUDY 2: AUTOMATED BALER

1.4.2.1 Description of Operation

The facility manufactures paper-packaging products. The intervention was performed on a waste-paper baling operation.

1.4.2.2 Hazard Identification

The waste-paper baling operation required three operators to spend approximately 30 min at the end of each shift (three shifts per day) loading paper scrap into the existing manually loaded scrap baler. The operation was labor-intensive, with operators grabbing armfuls of shredded paper and cramming the scrap into the baler, and required awkward lifting, twisting, and posturing. An ergonomics risk assessment determined that, because of the various and continual ergonomic stresses present, the operation posed a high risk of causing serious musculoskeletal injuries.

The facility had not experienced any ergonomic injuries associated with the baling operation, but from past company records the medical and disability costs associated with lumbar injuries averaged from $7500 to $50,000 per injury.

1.4.2.3 Hazard Intervention

The company decided to eliminate the hazard by purchasing an automated baler to fully replace the manual handling associated with managing shredded paper scraps.

1.4.2.4 Impacts of the Intervention

The intervention completely eliminated the risks associated with the manual handling during the waste baling operation. The new baler takes waste directly from the packaging production equipment and automatically bales and stacks it. In addition to removing the ergonomic risks, the intervention eliminated the need for three operators to spend 30 min at the end of three daily shifts to hand-load scrap onto the old baler. The intervention also eliminated the need for operators to wear PPE to avoid eye hazards and dust. The intervention also raised operator morale by eliminating an unpopular task that was frequently rotated among the 47 production workers at the facility.

In addition to the direct labor-saving benefits, the automated baler also reduced the amount of paper dust generated during the scrap-handling operation. This resulted in less paper dust being distributed throughout the site, thus reducing cleaning space and saving labor time, also contributing to a cleaner process and product. The property insurance provider considered the reduction in dust build-up to have lowered the facility's fire risk.

1.4.2.5   Financial Metrics

The five-year NPV of the project was −$1385 using a discount rate of 8% and an inflation rate of 3%. The only costs included in the analysis were the labor savings from eliminating the need to manually load the baler and the capital cost of the baler purchase and installation. The costs associated with injury reduction, facility cleaning, and PPE elimination were not included.

1.4.2.6   Lessons Learned

Although the project did not yield a sizable financial return on investment, the intervention did return the company's cost of capital while reducing a significant risk of injury due to manual handling. The project also illustrated that improvement in health and safety conditions often results in improved labor productivity. In this case, the positive benefits of the intervention were transferable to other facilities within the company, thus serving as good practice for the corporation.

1.4.3   CASE STUDY 3: CARBON MONOXIDE CONTROL

1.4.3.1   Background

The following case study focuses on a heat-treating facility in a company with operations in industrial manufacturing. The heat-treating process entailed open-room exhaust of natural gas-fired furnaces and open-room exhaust of endogas (a carbon-rich atmosphere used in heat-treating furnaces). The only ventilation was achieved through axial roof fans.

1.4.3.2   Hazard Identification

The hazard identified with this particular industrial manufacturing operation involved carbon monoxide (CO) exposure to employees working within a heat-treating facility. CO is an odorless, colorless, and tasteless poisonous gas. CO is harmful when inhaled because it displaces oxygen in the blood and deprives vital organs such as the heart and brain from receiving oxygen. CO poisoning can be reversed if caught in time, but, even with recovery, acute poisoning may cause permanent damage. Occupational Safety and Health Administration (OSHA) standards prohibit worker exposure to more than 50 ppm over an 8-h time-weighted average (TWA).

1.4.3.3   Hazard Intervention

The company identified the hazard as a chemical exposure to employees. Abatement of the hazard involved a change in the administrative and engineering controls. Data points for CO were routinely collected and administrative controls were implemented as necessary. The corporate goal for CO levels was less than half the threshold limit value (TLV) for CO (12.5 ppm). This goal was reached by implementing local exhaust ventilation (LEV) as the primary engineering control. All CO emission points (burner and endogas exhausts) were identified and targeted for LEV source controls. A ventilation system with variable-speed fans controlled by real-time direct reading electrochemical sensors for CO was installed in the heat-treating facility.

1.4.3.4   Impacts of the Intervention

There were many positive health, business, and risk-management results from the implementation of the engineering controls. Health improvements resulted from the intervention because employees were not directly exposed to CO, and were healthier, happier, and more comfortable in the workplace. Health-related absenteeism was reduced drastically. Employee morale increased significantly, improving the quality of work. The business process was improved because there was a reduction of CO concentration in the heat-treating facility.

While this project did not demonstrate a significant financial payback, many other benefits resulted from it. The project demonstrated leadership commitment to health service executives. Aesthetic improvement in the facility resulted from the use of a state-permitted LEV system to remove smoke and haze. There were no changes in product quality or customer satisfaction or service resulting from the intervention.

1.4.3.5   Financial Metrics

The project's capital requirements were $1.6 million to install the ventilation system. The intervention resulted in a negative NPV of −$1,005,597. The internal rate of return (IRR) was −25% while the return on investment (ROI) was −56%. Utility costs associated with running the IH-related equipment were expected to increase once the intervention was in place.

1.4.3.6   Lessons Learned

Retrospective analyses do not provide the opportunity to evaluate the costs and benefits of alternative hazard control solutions, but even in negative cost situations, IH value can be demonstrated. In this case, the heat-treating operation was an ultimate financial negative but a health, morale, and productivity positive. The benefits were valuable to the management, and in time will likely show financial payback as well.

1.4.4   Case Study 4: Metal Removal Fluid Management Control Plan

1.4.4.1   Description of Operation

The following case study involves a global transportation company with operations in auto manufacturing. This case study focuses on a machining department and the processes involved in an automotive transmission machining plant where metal removal fluids (MRFs) such as lubricants and coolants are utilized in production processes.

1.4.4.2 Hazard Identification

The hazard identified with this particular equipment manufacturing operation involved employee exposure to contaminated MRFs in the automotive transmission machining plant. In 2002, an employee reported to the plant medical department with complaints of respiratory illness while working on the machining plant where MRFs such as lubricants and coolants were utilized in production processes.

A subsequent medical examination confirmed that the employee was diagnosed with occupational hypersensitivity pneumonitis (HP). The employee received medical treatment and was placed on medical leave, and an investigation of the cause of the disease was undertaken. HP is a serious lung disease associated with exposure to microbiologically contaminated aerosols of synthetic, semi-synthetic, and soluble oil metalworking fluids. In the short term, HP is characterized by coughing, shortness of breath, and flu-like symptoms (fevers, chills, muscle aches, and fatigue). The chronic phase (following repeated exposures) is characterized by lung scarring associated with permanent lung damage.

1.4.4.3 Hazard Intervention

The company identified the hazard as microbiological contamination of the MRF. Abatement involved changing the type of fluid in use and implementing a comprehensive MRF control plan that provided for proper selection of MRFs, development of efficient coolant and machine maintenance schedules, and design of effective ventilation systems to maximize control of coolant aerosols.

The initial study and completion of IH risk assessments did not identify a clear relationship between known air contaminants in the work environment and the respiratory disease. Therefore, a multifunctional task force was created, with the primary objective to eliminate the risk of respiratory disease (HP) associated with MRFs. The task force had representation from the following groups: division and plant functions, corporate/plant IH, corporate research and development IH, plant union safety and health and IH, division/plant medical, corporate/plant environmental engineering and chemical management, plant manufacturing leadership, manufacturing engineering, and maintenance. The task force conducted numerous exposure assessments, research studies, production process changes, and maintenance process improvements.

1.4.4.4 Impacts of the Intervention

There were many positive health, business, and risk-management benefits that resulted from the implementation of the comprehensive MRF control plan. Health improvements resulted from the intervention because the air contaminant exposure associated with MRF machining was eliminated or reduced and employees were no longer directly exposed. No further cases of HP have been reported in the 4 years following the intervention. Employee respiratory complaints were eliminated or reduced. Employees were healthier, happier, and more comfortable in the workplace. Employee morale increased significantly, improving the trust and confidence of employees in the safety and health program.

The business process was improved as tooling life was extended and therefore tooling costs were reduced. Many risk-management benefits resulted from the intervention, including enhanced relationships between the division and plant union management. Management and engineering systems to support MRF safety and health goals were enhanced. Another benefit involved the development of improved bio-stable coolant strategies.

1.4.4.5 Financial Metrics

As part of the value study, a retrospective analysis was conducted with an incremental approach to reduce workplace illnesses and to improve risk management and business processes. Using the Value Study Data Collection Tool and entering the data in the ROHSEI software, the NPV calculated for the project duration of 5 years was $991,888. The IRR was 120%, while the ROI was 22%. The DPP was 0.5 years. Total costs after reducing, mitigating, or controlling the IH hazards were $2,883,573.

Management also realized that they needed to continue their efforts to reduce employee exposures to air contaminants from MRFs through a comprehensive MRF control plan.

1.4.4.6 Lessons Learned

Without IH involvement in this problem, it would have been difficult to identify the source of the hazard because the relationship between illness and MRF is not well understood. Experience with investigating complaints of this nature helped IH identify the microbiological nature of the hazard and make recommendations that solved the problem.

Ultimately, the task force concluded that an effective MRF management program is essential for ensuring the health and safety of employees working in aluminum and iron metal machining operations.

1.5 PtD AND SUSTAINABILITY

The concepts of sustainability and PtD were identified as very congruent and able to coexist [44]. PtD is linked to sustainability in many ways. Sustainability refers to accepting a duty to seek harmony with other people and with nature. Sustainability is not just about the environment. It is sharing with each other and caring for the Earth.

Figure 1.11 shows the interconnectivity of environment, economy, and society. As shown in this figure, sustainability is a multidimensional concept, involving environmental equity, economic equity and social equity. Therefore, an appropriate measurement framework should cover the economic, social, and environmental dimensions of sustainable development. As shown in this figure, ethics are the building blocks of sustainable development and should be incorporated into design development strategy to ensure long-term sustainability.

For example, a sustainable building project must not result in undesirable harm to the environment during its construction and use. The building must also make economic sense such that, in the long term, the revenues will at least equal the expenses of constructing and operating it. Finally, the building must be socially acceptable such that it will not cause any harm to any person or cause a group of people to experience injustice. What could be more unreasonable than to have workers construct a building that is not as safe to build as it could be? A fair construction project is when

FIGURE 1.11 Sustainability and interconnectivity of environment, economy, and society.

the designers have made reasonable effort to "design out" or minimize hazards and risks early in the design process. Sustainable construction occurs when design contributes to safety [58].

The engineering profession is being challenged with a new and forceful set of requirements: population growth, resource scarcity, and environmental change. These include apparent changes to the atmosphere, hydrosphere, and biosphere resulting in major shifts from the environmental norms under which the artifacts of our civilization were originally designed. At one time, these aspects of engineering design could be taken for granted, because of the obvious stability of the environment within a narrow, acceptable, and predictable range of change. The added interconnectivity and complexity of the environment, shifting requirements from environmental changes, cannot be easily addressed with methods developed in the industrial age.

1.5.1 TRANSDISCIPLINARY SUSTAINABLE DEVELOPMENT

Figure 1.12 shows a widely accepted concept of sustainable development—interconnectivity of environment, economy, and society. The environment plays an important role in the well-being of community development. It affects a broad range of social and economic variables that have a vital impact on the quality of community life, human health, and safety. A dynamic environment contributes to a healthier society and a stronger economy. Similarly, the environment is itself affected by economic and social factors.

Traditionally, development was strongly related to economic growth, which provides economic prosperity for members of society. During the early 1960s, the growing numbers of poor in developing countries resulted in considerable attempts to improve income distribution to the poor. As a result, the development paradigm changed toward equitable growth, where social (distributional) objectives, especially poverty alleviation, were accepted to be as important as economic efficiency. By the early 1980s, clear evidence proved that environmental degradation was a major barrier to development. Hence, protection of the environment became the third major element of sustainable development [59].

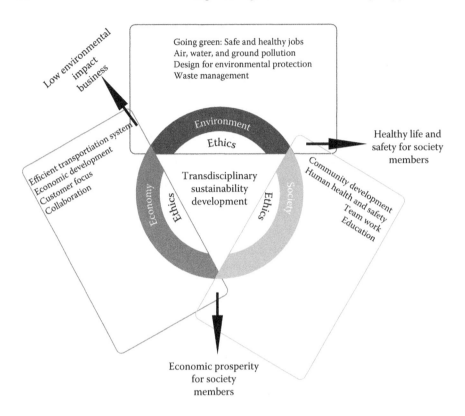

FIGURE 1.12   Transdisciplinary sustainable development.

1.5.2   Contaminated Environment

> Over increasingly large areas of the United States, spring now comes unheralded by
> the return of the birds, and the early mornings are strangely silent where once they
> were filled with the beauty of bird song.
>
> **Rachel Carson**

Rachel Carson combined her interests in biology and writing as a government
scientist with the Fish and Wildlife Service in Washington, DC. Her book titled
Silent Spring is credited with inspiring much of the late-twentieth century's environmental concern as she documented the effect of pesticides on the ecology.

> These sprays, dusts, and aerosols are now applied almost universally to farms, gar-
> dens, forests, and homes—nonselective chemicals that have the power to kill every
> insect, the "good" and the "bad," to still the song of birds and the leaping of fish in the
> streams, to coat the leaves with a deadly film, and to linger on in soil—all this though
> the intended target may be only a few weeds or insects. Can anyone believe it is pos-
> sible to lay down such a barrage of poisons on the surface of the earth without making
> it unfit for all life? They should not be called "insecticides," but "biocides."
>
> **Rachel Carson**

The condition of the environment and what can be done to protect it in the future rank high among the concerns of Americans in the twenty-first century. The environmental degradation that has occurred during the intervening years makes it devastatingly clear that continued growth in population and economic development make the correction of past ecological misuse complex and expensive. Hazardous substances, including chemicals, pesticides, heavy metals, and other toxic substances from industrial processes, refueling facilities, and agriculture, dumped at uncontrolled hazardous waste sites have been seeping into the ground and aquifer for several years. Scientists and engineers must begin to recognize the delicate nature of the environment in their endeavors and give it the priority it deserves.

1.5.2.1   Air Pollution

The quality of air that surrounds the earth has been degraded to the extent that warnings are issued in many cities when contamination levels reach the hazard zone. Joggers are warned about jogging at times of the day when smog levels are high, and many metropolitan cities in the world have enacted laws to control motor vehicle and other industrial emissions in an effort to lower air pollution levels. In Mexico City, more than 21 million people live in an atmosphere so foggy that the sun is obscured and so poisonous that school is sometimes delayed until late morning when the air clears. Air pollution can be prevented by lowering emission levels from motor vehicles and using more environment-friendly commercial products. Factories that produce hazardous air pollution should use "scrubbers" or other procedures on their smoke stacks to eliminate contaminants before they enter the air outside the plant.

1.5.2.2   Groundwater Contamination

Groundwater is one of the most essential natural resources and degradation of its quality has a major effect on the well-being of people. The quality of groundwater reflects inputs from the atmosphere, from soil and water–rock reactions, as well as from contaminant sources such as mining, land clearance, agriculture, acid precipitation, and industrial wastes. The fairly slow movement of water through the ground means that contaminants dwell longer in groundwater than in surface water. Groundwater is an important water resource that serves as a source of drinking water for the majority of people living in the United States. Contamination from natural and human sources can affect the use of groundwater. For example, spilling, leaking, improper disposal, or accidental and intentional application of chemicals on the land surface will result in overspill that contaminates nearby streams and lakes.

Strong competition among the agricultural, industrial, and domestic sectors is lowering the groundwater table. The quality of groundwater is severely affected because of the extensive pollution of surface water. The sustainability of groundwater utilization must be assessed from a transdisciplinary perspective, where hydrology, ecology, geomorphology, and climatology play important roles.

Environmental problems are essentially research and development challenges of a different order. These problems can be solved by scientists and engineers working together with political entities that can enact the necessary legislation, obtain the required international cooperation, and provide the necessary funding. The environment can no longer be considered a bottomless reservoir in which chemical discharges, toxic material, and harmful stack vapors can be deposited because of the lack of a measurable deleterious effect on the immediate surroundings.

Managing the environment is an international problem that cannot be based on monitoring and controlling at the local level only. Engineers and scientists must play key roles in providing the essential technology for understanding these global problems and in implementing workable solutions.

1.5.3 MAKING "GREEN JOBS" SAFE: INTEGRATING OCCUPATIONAL SAFETY AND HEALTH INTO GREEN JOBS AND SUSTAINABILITY

In 2008, the world experienced the worst financial crisis of our generation, triggering the start of the most difficult recession since the Great Depression. The financial crisis has forced policymakers to respond powerfully, creatively, and positively to resolve issues: interest rates have been considerably reduced, a stimulus package for a green economy has been signed, hundreds of billions of dollars have been provided to banking systems around the world, a stimulus package has been planned to create or save up to 3.6 million jobs over the next two years, to increase consumer spending, and to stop the recession.

Barbier suggested that an investment of 1% of global gross domestic product (GDP) over the next two years could provide the critical mass of green infrastructure needed to seed a significant greening of the global economy: "Green stimulus is well within the realm of the possible: at one percent of global GDP" [60,61].

Although many elements of the green economy have value-added benefits for a global economy, we should remain health conscious of the potential hazards that workers face when performing "green jobs."

Schulte and Heidel stated that:

> There are benefits as well as challenges as we move to a green economy. Defined broadly, green jobs are jobs that help to improve the environment. These jobs also create opportunities to help battle a sagging economy and get people back to work. Yet, with the heightened attention on green jobs and environmental sustainability, it is important to make sure that worker safety and health are not overlooked. NIOSH and its partners are developing a framework to create awareness, provide guidance, and address occupational safety and health issues associated with green jobs and sustainability efforts [62].

Figure 1.13 shows how our knowledge about old and new hazards intersects challenges created by new technologies and adaptations of work activities to perform green jobs [62].

Although many green job programs have set the commendable goal of recruiting young workers into the workforce, it is known that these inexperienced new workers could be the most at risk for job-related injuries. Moreover, in addition to these green job programs, stimulus package spending on infrastructure projects will also expose thousands of new workers to the myriad hazards encountered in the construction of bridges, highways, and public buildings. Hazards expected to be encountered in green jobs include [63]:

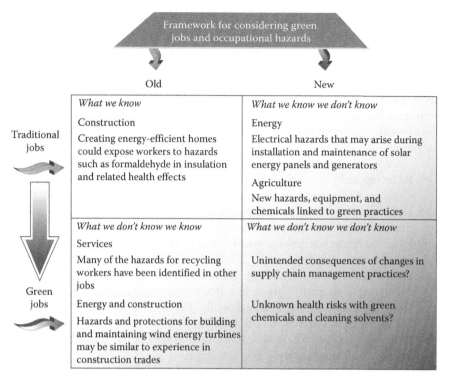

	Old	New
Traditional jobs	*What we know* Construction Creating energy-efficient homes could expose workers to hazards such as formaldehyde in insulation and related health effects	*What we know we don't know* Energy Electrical hazards that may arise during installation and maintenance of solar energy panels and generators Agriculture New hazards, equipment, and chemicals linked to green practices
Green jobs	*What we don't know we know* Services Many of the hazards for recycling workers have been identified in other jobs Energy and construction Hazards and protections for building and maintaining wind energy turbines may be similar to experience in construction trades	*What we don't know we don't know* Unintended consequences of changes in supply chain management practices? Unknown health risks with green chemicals and cleaning solvents?

FIGURE 1.13   Framework for considering green jobs and occupational hazards.

- Exposure to lead and asbestos in the course of energy-efficient retrofitting and weatherization in older buildings
- Respiratory hazards from exposure to fiberglass and other materials in reinsulation projects
- Exposure to biological hazards, such as molds, in fixing leaks
- Crystalline silica exposure from fiber-cement materials, which may contain up to 50% silica
- Ergonomic hazards from installation of large insulation panels
- Fall hazards during the installation of heavy energy-efficient windows and solar panels and during the construction and maintenance of windmills (typically 265 ft tall)
- Electrical hazards encountered in the course of weatherization projects

Green initiatives like recycling can have amazing success. However, that does not automatically imply they are good for the earth, for society, or for those working in green jobs. For example, more than 50% of refined lead is now produced from recycled material. On the contrary, global lead production has increased considerably since 2003, placing a new generation at risk from an old and very toxic hazard.

To give another example, solar energy will play an essential role in meeting challenges such as human energy needs and global warming, in reducing U.S. dependence on energy imports, in creating green jobs, and in helping revitalize the U.S. economy. However, as the photovoltaic (PV) solar panel sector expands rapidly, little attention is being paid to the possible environmental and health costs. The most commonly used PV solar panels are based on materials and processes from the microelectronics industry and have the capability to create a huge new wave of electronic waste (e-waste) at the end of their useful lives. Recommendations to build a safe and sustainable solar energy industry include [63]:

- Reducing and eventually eliminating the use of toxic materials and developing environmentally sustainable practices
- Ensuring that PV solar manufacturers are responsible for the lifecycle impacts of their products through extended producer responsibility (EPR)
- Ensuring proper testing of new and emerging materials and processes based on a precautionary approach
- Expanding recycling technology and design products for easy recycling
- Promoting high-quality green jobs that protect worker health and safety and provide a living wage throughout the global PV solar industry, including supply chains and end-of-life recycling
- Protecting community health and safety throughout the global PV solar industry, including supply chains and recycling

1.5.4 GOING GREEN DURING CONSTRUCTION

Going green during the construction assures the benefit of the surrounding community, workers, and visitors on the site by reducing emissions, airborne pollution, and toxic gases like CO.

Green building development focuses on energy efficiency and on using less toxic products from the perspective of future occupants of a building It also includes addressing air quality issues, such as diesel exhaust generated by vehicles (which contains nitrogen oxides, sulfur oxides, and polycyclic aromatic hydrocarbons), which in turn increases the risk of lung and perhaps bladder cancer and includes other health problems such as asthma and cardiovascular diseases. Similar problems can be expected from gasoline-powered vehicles as well.

Dust is another issue in air quality. Dust consists of small solid particles created by a breakdown or fracture process, such as grinding, crushing, or impact. Particles that are too large to stay airborne settle while others remain in the air indefinitely. General dust levels at considerably high concentrations may induce permanent changes to airways and loss of functional lung capacity.

Silica dust is accountable for a major American industrial disaster. Every year about 300 workers die from silicosis, a chronic disabling lung disease caused by the formation of nodules of scar tissue in the lungs. Hundreds more are disabled and between 3000 and 7000 new cases occur each year. High-risk work activities in construction include [64]:

- Chipping, drilling, and crushing rock
- Abrasive blasting
- Sawing, drilling, grinding, concrete and masonry products containing silica
- Demonstration of concrete/masonry
- Removing paint and rust with power equipment
- Dry sweeping or air blowing of concrete rock sand dust
- Jack hammering on concrete, masonry, and other surfaces

Detailed information on this subject can be found in Ref. [64].

1.6 CONCLUSION

It should be obvious that the material presented in this chapter constitutes only cursory treatment of the very broad and important subject of the prevention of occupational injuries, illnesses, and fatalities, in order to anticipate and "design-out" or minimize hazards and risks when new equipments, processes, and business practices are developed. However, some understanding of the relative role of the PtD process is important; thus, a brief concept description of the transdisciplinary PtD process has been included in this report.

REFERENCES

1. Luszki, M. B. 1958. *Interdisciplinary Team Research: Methods and Problems. Vol. 3, Research Training Series*, National Training Laboratories, National Education Association, New York University Press, New York, p. 304.
2. Nicolescu, B. 2002. *Manifesto of Transdisciplinarity.* Translated from French by K. C. Voss, State University of New York Press, Albany, 43pp.
3. Stokols, D. 1998. *The Future of Interdisciplinarity in the School of Social Ecology.* School of Social Ecology, University of California, Irvine. http//eee.uci.edu/98f/50990/readings.htm
4. Upham, S. 2001. *The Aims of Graduate Education*, Keynote address to the Society for Design and Process Workshop on Global Transdisciplinary Education, Pasadena, CA.
5. Meilich, A. 2006. System of Systems (SoS) engineering & architecture challenges in a net centric environment. *IEEE/SMC International Conference on System of Systems Engineering*, pp. 5, ISBN: 1-4244-0188-7.
6. Pickett, J. P. 2000. *American Heritage Dictionary of the English Language.* "Discipline." Houghton Mifflin, Boston, MA.
7. *Merriam-Webster Online Dictionary.* 2009. *Merriam-Webster Online.* "Discipline."
8. Klein, J. T. 2004. Disciplinary origins and differences. *Fenner Conference on the Environment, Understanding the Population–Environment Debate: Bridging Disciplinary Divides.* Canberra, Australia, pp. 10 of 20. http://www.science.org.au/events/fenner/klein.htm. pg. 9.
9. Fowers, B. J. 2008. From continence to virtue: Recovering goodness, character unity, and character types for positive psychology. *Theory & Psychology*, 18(5), 629–653.
10. Senge, P. M. 2006. *The Fifth Discipline: The Art and Practice of the Learning Organization.* Random House, Inc., New York.
11. The Carnegie Foundation for the Advancement of Teaching. 2008. *About Us.* http://www.carnegiefoundation.org/index.asp

12. Sheppard, S. D., Macatangay, K., Colby, A., and Sullivan, W. N. 2009. *Educating Engineers, Designing for the Future of the Field.* Carnegie Foundation for the Advancement of Teaching, San Francisco: Jossey-Bass.
13. Petts, J., Owens, S., and Bulkeley, H. 2008. Crossing boundaries: Interdisciplinarity in context of urban environments. *Geoforum*, 39, 593–601.
14. Karen, C. 2008. *Transdisciplinary Research (TDR) and Sustainability Report*, Ministry of Research, Science and Technology (MoRST), Environmental Science and Research, Ltd., New Zealand.
15. Ertas, A., Maxwell, T. T., Tanik, M. M., and Rainey, V. 2003. Transformation of higher education: The transdisciplinary approach in engineering. *IEEE Transactions on Education*, 46(1), 289–295.
16. Gumus, B., Ertas, A., Tate, D., and Cicek, I. 2008. The transdisciplinary product development lifecycle (TPDL) model. *Journal of Engineering Design*, 19(03), 185–200.
17. Tate, D. Maxwell, T. T., Ertas, A., Zhang, H.-C., Flueckiger, P., Lawson, W., and Fontent, A. D. 2010. Transdisciplinary approaches for teaching and assessing sustainable design. *International Journal of Engineering Education*, 26(2), 1–12.
18. Kollman T. and Ertas, A. 2010. Defining transdisciplinarity. *The ATLAS T3 Bi-Annual Meeting Proceedings, Vol. 1.* The ATLAS Publications, USA, pp. 22–38.
19. Mittelstrass, J. 2003. *Transdisziplinarität* [Transdisciplinarity] [in German], wissenschaftliche Zukunft und institutionelle Wirklichkeit, Germany, ISBN: 387940786X.
20. Transdisciplinary Centers for Behavioral and Preventive Medicine. 2009. Training: Training the next generation of researchers, http://www.lifespan.org/behavmed/TTURCTrainingNext.htm
21. Nicolescu, B. 2005. *Toward Transdisciplinary Education and Learning. Science and Religion: Global Perspectives*, CNRS, University of Paris, pp. 1–12.
22. Hadorn, H. G., Biber-Klemm, S., Grossenbacher-Mansuy, W., Joye, D., Pohl, C., Wiesmann, U., and Zemp, E. 2008. Chapter 29—Enhancing transdisciplinary research: A synthesis in fifteen propositions. *Handbook of Transdisciplinary Research*, Springer, p. 435.
23. Peterson, L. C. and Martin, C. 2005. A new paradigm in general practice research towards transdisciplinary approaches. http://www.priory.com/fam/paradigm.htm
24. Stokols, D., Harvey, R., Gress, J., Fuqua, J., and Phillips, K. 2004. *In vivo* studies of transdisciplinary scientific collaboration, lessons learned and implications for active living research. *American Journal of Preventive Medicine*, 28(2), 202–213.
25. Petts, J., Owens, S., and Bulkeley, H. 2008. Crossing boundaries: Interdisciplinarity in context of urban environments. *Geoforum*, 39, 597.
26. Danish Wind Industry Association. 2009. Green Energy Ohio. http://www.greenenergy ohio.org/page.cfm?pageId=341
27. Wind turbine history, alternative-energy-resources. 2009. http://www.alternative-energy-resources.net/wind-turbine-history.html
28. National Institute for Occupational Safety and Health (NIOSH). 2007. *Prevention through Design in Motion*, February 5.
29. Fisher J. M. 2008. Healthcare and social assistance sector. *Journal of Safety Research*, 39, 181.
30. Schulte, P. A., Rinehart, R., Okun, A., Geraci, C. L., and Heidel, D. S. 2008. National Prevention through Design (PtD) initiative. *Journal of Safety Research*, 39, 120.
31. Ertas, A. and Jones, J. 1996. *The Engineering Design Process*, 2nd edition, John Wiley & Sons, Inc., New York.
32. U.S. Department of Energy. 2008. *DOE Standard, Integration of Safety into the Design Process*, Report, DOE-STD-1189-2008.
33. Gage, H. 1989. Integrating safety and health into M. E. Capstone design courses. *ASEE, Southwest Regional Conference*, Texas Tech University, Lubbock, TX.

34. Haddon, W. 1980. The basic strategies for reducing damage from hazards of all kinds. *Hazard Prevention*, 16, 8–12.

35. National Institute for Occupational Safety and Health (NIOSH). 2009. Prevention through Design. http://www.cdc.gov/niosh/topics/ptd/

36. Heidel, D. S. and Schulte, P. A. 2008. Making the business case for Prevention through Design. NIOSH Science Blog, February 6. http://www.cdc.gov/niosh/blog/nsb060208_ptd.html

37. All Business, AD&B Company. 2008. Prevention through Design. http://www.allbusiness.com/labor-employment/workplace-health-safety/11715914-1.html

38. National Institute for Occupational Safety and Health (NIOSH). 2008. Prevention through Design. http://www.cdc.gov/niosh/programs/ptdesign/

39. Schulte, P. A. and Heidel, D. S. 2009. Prevention through design. *Motion*, June 17.

40. Bell, C., Stout, N., Bender, T., Conroy, C., Crouse, W., and Myers, J. 1990. Fatal occupational injuries in the United States. *Journal of the American Medical Association*, 263(22), 3047–3050.

41. National Institutes of Health. 1993. Safety note number 8: Proper handling and disposal of sharps. http://www.niehs.nih.gov/odhsb/notes/note8.htm

42. National Institute for Occupational Safety and Health (NIOSH). 2008. Preventing back injuries in healthcare settings. http://www.cdc.gov/niosh/blog/nsb092208_lifting.html

43. Bealko, B. S., Kovalchik, G. P., and Matetic, J. R. 2008. Mining sector. *Journal of Safety Research*, 39, 187–189.

44. Behm, M. 2008. Construction sector. *Journal of Safety Research*, 39, 175–178.

45. Husberg, B. 2008. Agriculture, forestry, and fishing sector. *Journal of Safety Research*, 39, 171–173.

46. Fisher, J. M. 2008. Healthcare and social assistance sector. *Journal of Safety Research*, 39, 179–181.

47. Heidel, S. D. 2008. Manufacturing sector. *Journal of Safety Research*, 39, 183–186.

48. Johnson, V. J. 2008. Services sector. *Journal of Safety Research*, 39, 191–194.

49. Madar, A. S. 2008. Transportation, warehousing, and utilities sector. *Journal of Safety Research*, 39, 195–197.

50. Mroszczyk, J. W. 2008. Wholesale and retail trade sector. *Journal of Safety Research*, 39, 199–201.

51. Gambatese, A. J. 2008. Research issues in Prevention through Design. *Journal of Safety Research*, 39, 153–156.

52. Mann, A. J. 2008. Education issues in Prevention through Design. *Journal of Safety Research*, 39, 165–170.

53. Lin, M.-L. 2008. Practice issues in Prevention through Design. *Journal of Safety Research*, 39, 157–159.

54. Howe, J. 2008. Policy issues in Prevention through Design. *Journal of Safety Research*, 39, 161–163.

55. U.S. Department of Labor. 2006. *Injuries, Illnesses, and Fatalities. Industry Injury and Illness Data*. U.S. Department of Labor, Bureau of Labor Statistics, Safety and Health Statistics Program, Washington, DC.

56. American Industrial Hygiene Association. 2008. Strategy to demonstrate the value of industrial hygiene. http://www.aiha.org/votp/AIHA_4.html

57. American Industrial Hygiene Association. 2009. IH value. http://www.aiha.org/votp_NEW/index.html

58. Prevention through Design: Design for Construction Safety. 2009. http://www.designforconstructionsafety.org/concept.shtml

59. Munasinghe, M. 2007. Sustainomics and sustainable development. *The Encyclopedia of Earth*, May 7. http://www.eoearth.org/article/Sustainomics_and_sustainable_development

60. *Global Green New Deal, Policy Brief.* 2009. Published by the United Nations Environment Programme as part of its Green Economy Initiative, in collaboration with a wide range of international partners and experts.
61. Barbier, E. B. 2009. *A Global Green New Deal.* UNEP-DTIE, February.
62. Schulte, P. A. and Heidel, S. 2009. Going green: Safe and healthy jobs. *Prevention through Design in Motion*, 5(July).
63. National Council for Occupational Safety and Health. 2009. Green jobs—Safe jobs campaign. http://coshnetwork6.mayfirst.org/node/133
64. 2009. Toward a just and sustainable solar energy industry. *Silicon Valley Toxics Coalition White Paper*, January 14. http://www.coshnetwork.org/node/266
65. Celenza, J. 2009. Green during Construction Project. RI Committee on Occupational Safety and Health.

2 Static

2.1 INTRODUCTION

Mechanics can be defined as the science that describes and predicts the condition of rest or the motion of bodies under the action of applied forces. Static analysis is used when the body is in equilibrium (at rest) and is subjected to a force system. Static analysis is the study of centroids, the center of gravity, and moment of inertia of free bodies. It also includes finding the resultant forces and moments acting on a free body and depicting these results diagrammatically.

2.2 FUNDAMENTAL CONCEPTS

2.2.1 WEIGHT AND MASS

The weight of a body is defined as the force exerted by the body due to its gravitational interaction with the earth. The mass, m, of a body is given by

$$m = \frac{W}{g} \tag{2.1}$$

where g is the acceleration of gravity ($g = 32.2$ ft/s^2, $g = 9.81$ m/s^2) and W is the weight of the body. The mass is that property of a body that measures its resistance to a change of motion.

2.2.2 RIGID BODY

A body is said to be rigid if the points on a body are fixed in a position relative to one another under the action of applied forces. Although solid bodies are never normally rigid, deformation under the action of applied forces is so small that it can be neglected in static analysis.

2.2.3 FORCE

A force is the action exerted by one rigid body upon another, which produces some mechanical effect. The most primitive notion of force is that of a push or pull action, as shown in Figure 2.1.

2.3 FORCE AS A VECTOR

Forces are often treated as vectors. A *vector* is a variable that has two properties: magnitude and direction. A variable that has magnitude and no direction is called *scalar*.

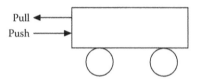

FIGURE 2.1   Pull and push forces acting on a cart.

A vector is represented graphically in Figure 2.2, where the length of the arrow represents the magnitude (5 U) of the vector and the angle α shows the direction of the vector. If $\bar{\lambda}$ represents a unit vector, force $\bar{F}$ can be written as

$$\bar{F} = F\bar{\lambda} = 5\bar{\lambda} \tag{2.2}$$

where F is the magnitude of vector $\bar{F}$. In two dimensions, force $\bar{F}$ can be broken into two vector components as shown in Figure 2.3.

$$\bar{F} = F_x\bar{i} + F_y\bar{j} \tag{2.3}$$

Unit vectors $\bar{i}$ and $\bar{j}$ have a magnitude of 1 and show x and y directions, respectively. If projection of $\bar{F}$ on x is $F_x = 3$ and projection of $\bar{F}$ on y is $F_y = 4$, then the force vector can be written as

$$\bar{F} = 3\bar{i} + 4\bar{j} \tag{2.4}$$

where the magnitude of $\bar{F}$ is given by

$$F = \sqrt{F_x^2 + F_y^2} = \sqrt{3^2 + 4^2} = 5 \tag{2.5}$$

Angle α between force F and axis x can be determined by

$$\cos\alpha = \frac{F_x}{F} \quad \text{or} \quad \sin\alpha = \frac{F_y}{F} \tag{2.6}$$

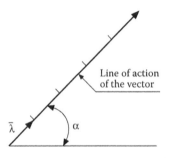

FIGURE 2.2   Graphical representation of a vector.

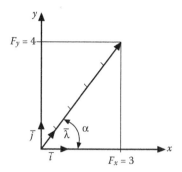

FIGURE 2.3 Vector components.

2.3.1 DEFINITION

Two vectors are said to be equal if they have the same magnitude and direction (see Figure 2.4). If the vectors are in the same plane, the sum, R, of the vectors can be obtained by using the parallelogram law or tip-to-tail addition as shown in Figure 2.5.

The resultant vector is $\bar{R} = \bar{F_1} + \bar{F_2}$.

2.3.2 TRIGONOMETRIC SOLUTION

a. *Sine law*: Relationships between forces and angles can be written as (see Figure 2.6)

$$\frac{F_1}{\sin \beta} = \frac{F_2}{\sin \alpha} = \frac{F_3}{\sin \theta} \tag{2.7}$$

FIGURE 2.4 Equal vectors.

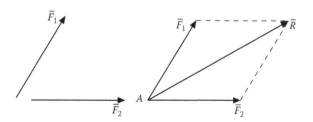

FIGURE 2.5 Sum of vectors in the same plane.

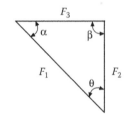

FIGURE 2.6 Sine law.

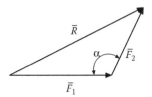

FIGURE 2.7 Cosine law.

b. *Cosine law*: Relationships between forces and the angle can be written as (see Figure 2.7)

$$R^2 = F_1^2 + F_2^2 - 2F_1F_2 \cos\alpha \qquad (2.8)$$

Example 2.1

A boat is being towed in a narrow channel by two boys pulling on ropes attached to the boat, as shown in Figure 2.8. If each boy applies a force of 10 N, determine the magnitude of the resultant force.

SOLUTION

Using the cosine law and Figures 2.8 and 2.9, the resultant force, R, is

$$R^2 = F_1^2 + F_2^2 - 2F_1F_2 \cos\theta$$

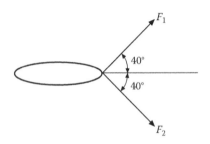

FIGURE 2.8 Towing boat.

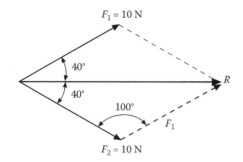

FIGURE 2.9   Force components.

Substituting known values yields

$$R^2 = 10^2 + 10^2 - 2 \times 10 \times 10 \times \cos 100° = 234.72$$

$$R = 15.32 \text{ N}$$

Example 2.2

Determine the magnitude and direction of force F_1 shown in Figure 2.10.

SOLUTION

From Figure 2.10, we have $R = 9000$ N and $\beta = 110°$.
  Applying the cosine law to determine the magnitude of F_1 gives

$$F_1^2 = R^2 + 2250^2 - 2R \times 2250 \times \cos\beta$$

Substituting known values yields

$$F_1^2 = 9000^2 + 2250^2 - 2 \times 9000 \times 2250 \times \cos 110° = 999.135 \times 10^5 \text{ N}$$

$$F_1 = 9995.7 \text{ N}$$

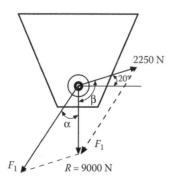

FIGURE 2.10   Applied force components.

Applying the sine law to determine the direction of F_1 gives

$$\frac{2250}{\sin \alpha} = \frac{9995.9}{\sin 110°}$$

The result of the above equation yields: $\alpha = 12.2°$.

2.4 FORCE COMPONENTS IN SPACE

To determine the rectangular components of a force in space (in three dimensions), consider Figure 2.11 showing a force vector $\overline{F}$ whose direction is specified by a unit vector $\overline{\lambda}$.

Assume that angles between $\overline{F}$ and x, y, and z are θ_x, θ_y, and θ_z, respectively. Components of $\overline{F}$ on the x, y, and z axis are given by

$$\begin{aligned} F_x &= F_h \times \cos\varphi \\ F_y &= F \times \cos\theta_y \\ F_z &= F_h \times \sin\varphi \end{aligned} \tag{2.9}$$

where F_h is the projection of $\overline{F}$ on the xz plane and is given as

$$F_h = F \times \sin\theta_y \tag{2.10}$$

Force vector $\overline{F}$ can also be written as

$$\overline{F} = F\overline{\lambda} \tag{2.11}$$

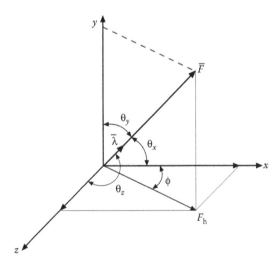

FIGURE 2.11 Rectangular components of a force vector.

where unit vector $\overline{\lambda}$ is defined as

$$\overline{\lambda} = \cos\theta_x\overline{i} + \cos\theta_y\overline{j} + \cos\theta_z\overline{k} \qquad (2.12)$$

Substituting Equation 2.12 into Equation 2.11 yields

$$\overline{F} = F\cos\theta_x\overline{i} + F\cos\theta_y\overline{j} + F\cos\theta_z\overline{k} \qquad (2.13)$$

The cosines of the angles θ_x, θ_y, and θ_z are called direction cosines of force vector $\overline{F}$. Since force components on the x, y, and z axis are given by

$$\begin{aligned} F_x &= F\cos\theta_x \\ F_y &= F\cos\theta_y \\ F_z &= F\cos\theta_z \end{aligned} \qquad (2.14)$$

Equation 2.13 can be written as

$$\overline{F} = F_x i + F_y j + F_z k \qquad (2.15)$$

The magnitude of force vector $\overline{F}$ is

$$F = \sqrt{F_x^2 + F_y^2 + F_z^2} \qquad (2.16)$$

Note that

$$\cos^2\theta_x + \cos^2\theta_y + \cos^2\theta_z = 1 \qquad (2.17)$$

The relationship given by Equation 2.17 enables us to determine the third angle if any two angles have been specified. Another useful relationship to determine direction cosines is

$$\frac{F_x}{\cos\theta_x} = \frac{F_y}{\cos\theta_y} = \frac{F_z}{\cos\theta_z} = \frac{F}{1} \qquad (2.18)$$

Example 2.3

As shown in Figure 2.12, the angle between a vertical boom AB and guy wire AC is 25°. If the tension in AC is 400 N, calculate:

a. x, y, and z components of the force exerted on the vertical boom at point B.
b. The angles θ_x, θ_y, and θ_z.

SOLUTION

Force components at point B are:

$$F_h = F_{AC} \times \sin 25° = 400 \times \sin 25° = 169\,\text{N}$$
$$F_x = F_h \times \cos \varphi = 169 \times \cos 30° = 146.4\,\text{N}$$

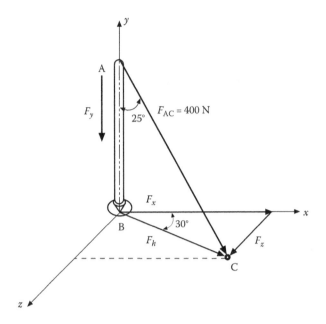

FIGURE 2.12   Vertical boom system.

$$F_z = F_h \times \sin \varphi = 169 \times \sin 30° = 84.5 \, \text{N}$$

$$F_y = F_{AC} \times \cos 25° = 400 \times \cos 25° = -362.5 \, \text{N}$$

Direction cosines are:

$$\cos \theta_x = \frac{F_x}{F} = \frac{146.4}{400} = 68.5°$$

$$\cos \theta_y = \frac{F_y}{F} = \frac{-362.5}{400} = 155°$$

$$\cos \theta_z = \frac{F_z}{F} = \frac{84.5}{400} = 77.8°$$

2.4.1   FORCE VECTOR DEFINED BY ITS MAGNITUDE AND TWO POINTS ON ITS LINE OF ACTION

As shown in Figure 2.13, a force vector can be defined by the coordinates of two points: M(x, y, z) and N(x, y, z). Vector representation of MN is given by

$$\overline{MN} = dx\,\overline{i} + dy\,\overline{j} + dz\,\overline{k} \qquad (2.19)$$

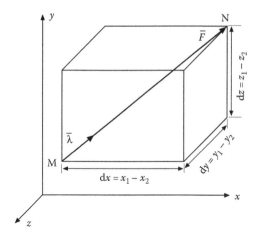

FIGURE 2.13 Defining force vector by two points.

The magnitude of MN is

$$MN = d = \sqrt{dx^2 + dy^2 + dz^2} \tag{2.20}$$

The unit vector on line MN is

$$\bar{\lambda} = \frac{\overline{MN}}{MN} = \frac{dx\bar{i} + dy\bar{j} + dz\bar{k}}{d} \tag{2.21}$$

Note that $\bar{\lambda}$ is also a unit vector on $\bar{F}$. The force vector can be written as

$$\bar{F} = F\bar{\lambda} = \frac{F}{d}(dx\,\bar{i} + dy\,\bar{j} + dz\,\bar{k}) \tag{2.22}$$

Defining

$$\cos\theta_x = \frac{dx}{d}$$

$$\cos\theta_y = \frac{dy}{d} \tag{2.23}$$

$$\cos\theta_z = \frac{dz}{d}$$

a force vector equation identical to Equation 2.13 can thus be obtained:

$$\bar{F} = F\cos\theta_x\bar{i} + F\cos\theta_y\bar{j} + F\cos\theta_z\bar{k} \tag{2.24}$$

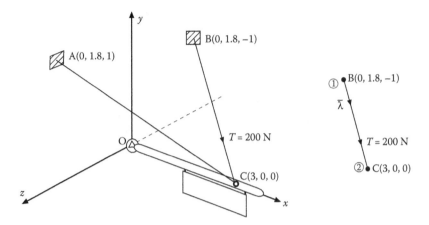

FIGURE 2.14 Cable system (unit for dimensions is meter).

Example 2.4

Assuming the magnitude of the tension in cable CB, shown in Figure 2.14, is 200 N, determine the tension T exerted on the anchor at B and direction cosines.

SOLUTION

The unit vector $\bar{\lambda}$ is

$$\bar{\lambda} = \frac{dx\, i + dy\, j + dz\, k}{d} = \frac{(3-0)i + (0-1.8)j + (0+1)k}{\sqrt{3^2 + 1.8^2 + 1^2}}$$

$$\bar{\lambda} = \frac{3i - 1.8j + k}{3.64}$$

The tension exerted on the anchor at B is

$$\bar{T} = T\bar{\lambda} = 200 \times \frac{3i - 1.8j + k}{3.64}$$

$$\bar{T} = 164.8i - 98.9j + 54.9k \text{ N}$$

From the unit vector components, direction cosines can be obtained as follows:

$$\cos\theta_x = \frac{dx}{d} = \frac{3}{3.64} \Rightarrow \theta_x = 34.5°$$

$$\cos\theta_y = \frac{dy}{d} = \frac{-1.8}{3.64} \Rightarrow \theta_y = 119.6°$$

$$\cos\theta_z = \frac{dz}{d} = \frac{1}{3.64} \Rightarrow \theta_z = 74°$$

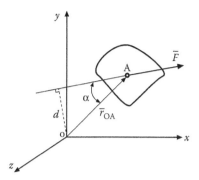

FIGURE 2.15 Rigid body.

2.5 MOMENTS

The moment of force $\overline{F}$ about any arbitrary point on a rigid body, shown in Figure 2.15, is the vector product of position vector $\overline{r}$ and $\overline{F}$:

$$\overline{M}_O = \overline{r}_{OA} \times \overline{F} \tag{2.25}$$

where $\overline{r}_{OA}$ is the position vector that starts from the point of interest and ends at the point at which the force is applied. The magnitude of the moment is

$$M = d \times F \tag{2.26}$$

where

$$d = r \times \sin \alpha \tag{2.27}$$

in three dimensions, the position vector $\overline{r}_{OA}$ is expressed as

$$\overline{r}_{OA} = r_x i + r_y j + r_z k \tag{2.28}$$

and the force $\overline{F}$ is

$$\overline{F} = F_x i + F_y j + F_z k \tag{2.29}$$

2.6 FREE-BODY DIAGRAM

Free-body diagram (FBD) is one of the most important concepts in mechanics, defined as a sketch of the isolated body (all the supports are removed and replaced by reaction forces) illustrating only the applied forces acting upon it. The attached weight is considered as the applied force and it must be shown in the FBD. For example, consider the beam–cable system shown in Figure 2.16a; an FBD of the system is shown in Figure 2.16b. Table 2.1 shows replacement reaction forces for the supports at A and B.

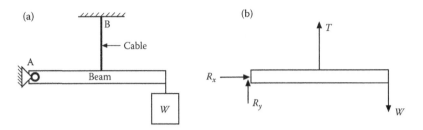

FIGURE 2.16   (a) Beam supported by cable and (b) free-body diagram.

TABLE 2.1

Examples of Replacement Reaction Forces to be Used for Different Types of Supports

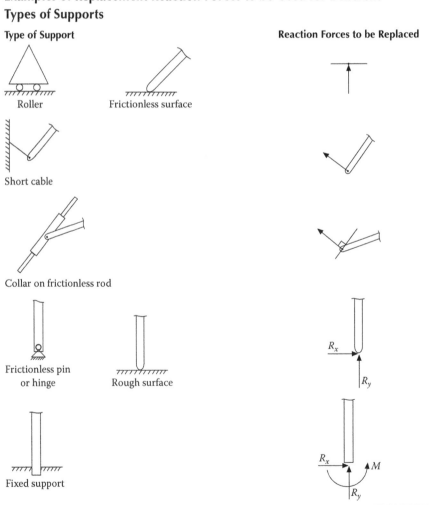

2.6.1 Equilibrium of a Particle

When the resultant of all forces acting on a particle is zero, then the particle is said to be in equilibrium. The equation for equilibrium is

$$\bar{R} = \sum \bar{F} = 0 \quad \text{or} \quad \sum F_x = 0; \quad \sum F_y = 0; \quad \sum F_z = 0 \qquad (2.30)$$

2.6.2 Equilibrium of a Rigid Body

A rigid body is said to be in equilibrium when

$$\bar{R} = \sum \bar{F} = 0 \quad \text{and} \quad \sum \bar{M}_A = 0 \qquad (2.31)$$

or

$$\begin{aligned}
\sum F_x = 0; \quad & \sum M_x = 0 \\
\sum F_y = 0; \quad & \sum M_y = 0 \\
\sum F_z = 0; \quad & \sum M_z = 0
\end{aligned} \qquad (2.32)$$

For rigid-body equilibrium, as shown in Equation 2.32, summation of the forces acting on the rigid body must be equal to zero. Also, summation of the moment components about any point "O" on the rigid body must be equal to zero. Note that both equations must be simultaneously satisfied. Therefore,

$$\begin{aligned}
\bar{M}_O &= \bar{r}_{OA} \times \bar{F} = 0 \\
\bar{M}_O &= (r_x i + r_y j + r_z k) \times (F_x i + F_y j + F_z k) = 0 \\
&= (r_y F_z - r_z F_y)i + (r_z F_x - r_x F_z)j + (r_x F_y - r_y F_x)k = 0
\end{aligned} \qquad (2.33)$$

Hence, for equilibrium, the components of the moment about any point can be written as

$$\begin{aligned}
M_x &= (r_y F_z - r_z F_y) = 0 \\
M_y &= (r_z F_x - r_x F_x) = 0 \\
M_z &= (r_x F_y - r_y F_x) = 0
\end{aligned} \qquad (2.34)$$

The moment components M_x, M_y, and M_z show the turning effect of force $\bar{F}$ about the x, y, and z axis, respectively.

2.6.3 Equilibrium of a Two-Force Member

A rigid body in equilibrium under the action of two forces is called a two-force member. These two forces must have the same magnitude, opposite directions, and the same line of action, as shown in Figure 2.17.

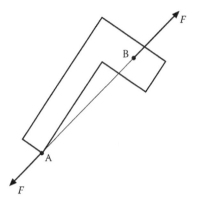

FIGURE 2.17   Two-force member.

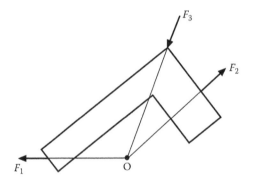

FIGURE 2.18   Three-force member.

2.6.4   EQUILIBRIUM OF A THREE-FORCE MEMBER

A rigid body in equilibrium under the action of three forces is called a three-force member. The line of action of these forces must intersect at one point, as shown in Figure 2.18.

2.6.5   EQUILIBRIUM OF A PULLEY SYSTEM

2.6.5.1   Simple Pulley

Tension in the rope is the same on each side of the pulley, as shown in Figure 2.19a. From the FBD shown in Figure 2.19b, for equilibrium of the system, T must be equal to $W/2$ as shown below:

$$\sum F_y = 0$$
$$T + T - W = 0$$
$$T = \frac{W}{2}$$

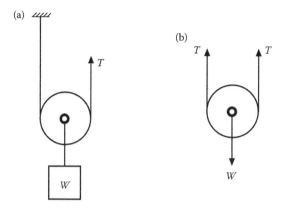

FIGURE 2.19   (a) Simple pulley and (b) free-body diagram.

2.6.5.2   Fixed Pulley

A fixed pulley, as shown in Figure 2.20a, only changes the direction of tension T. The same equilibrium equations can be obtained using the FBD shown in Figure 2.20b:

$$\sum F_y = 0$$
$$T + T - W = 0$$
$$T = \frac{W}{2}$$

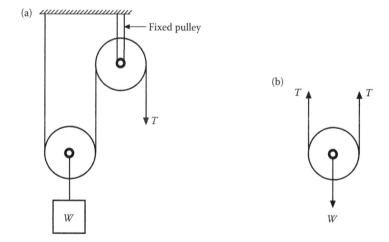

FIGURE 2.20   (a) Fixed pulley and (b) free-body diagram.

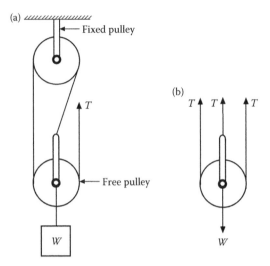

FIGURE 2.21 (a) Fixed- and free-pulley arrangement and (b) free-body diagram.

2.6.5.3 Fixed- and Free-Pulley Arrangement

To obtain the FBD of a fixed- and free-pulley arrangement, as shown in Figure 2.21a, a fixed or free pulley can be used. For example, to obtain the FBD of a free-pulley arrangement, all the ropes must be cut and must have tension in them, as shown in Figure 2.21b. From the FBD shown in Figure 2.21b, tension T can be calculated as follows:

$$\sum F_y = 0$$
$$T + T + T - W = 0$$
$$T = \frac{W}{3}$$

Example 2.5

Compute the reaction forces supported by the frictionless pin at point A of the frame loaded by using a two-force member (see Figure 2.22a).

SOLUTION

Dismember the structure shown in Figure 2.22a and draw two FBDs for CB and AB. Note that CB is a two-force member and the direction of the forces at C and B can be found from the FBD shown in Figure 2.22b. Next, draw an FBD to show the force at B acting on AB, as shown in Figure 2.22c. Using Figure 2.22c and

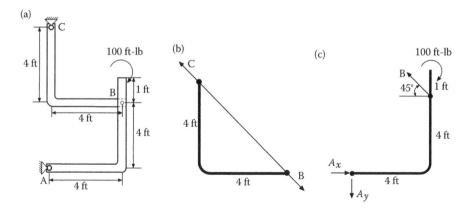

FIGURE 2.22 (a) Frame loaded by external force, (b) two force member, and (c) free-body diagram.

applying the equilibrium equation yields (assuming counterclockwise moments are positive)

$$\sum M_A = 0$$

$$\sum M_A = -100 + B \sin 45° \times 4 + B \cos 45° \times 4 = 0$$

$$5.66B = 100 \Rightarrow B = 17.67 \text{ lb}$$

From the FBD shown in Figure 2.22c

$$\sum F_x = 0$$

$$A_x - B \times \cos 45° = 0 \Rightarrow A_x = 17.67 \times 0.707 = 12.5 \text{ lb}$$

and

$$\sum F_y = 0$$

$$-A_y + B \sin 45° = 0 \Rightarrow A_y = 17.67 \times 0.707 = 12.5 \text{ lb}$$

Magnitude of force A is

$$A = \sqrt{A_x^2 + A_y^2} = \sqrt{12.5^2 + 12.5^2} = 17.68 \text{ lb}$$

Example 2.6

A force of 1.5 kN is applied at point A of the pipe shown in Figure 2.23. Determine the moment about point O.

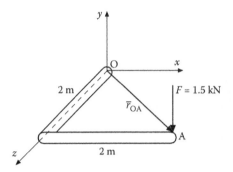

FIGURE 2.23   Pipe loaded by external force.

SOLUTION

The moment about point O is

$$\bar{M}_O = \bar{r}_{OA} \times \bar{F}$$

where position vector $\bar{r}_{OA}$ is

$$\bar{r}_{OA} = 2i + 2k$$

and

$$\bar{F} = -1.5j$$

Then

$$\bar{M}_O = (2i + 2k) \times (-1.5j)$$
$$\bar{M}_O = -3k + 3i \text{ kN m}$$

and the magnitude of the moment at point O is

$$\bar{M}_O = \sqrt{(-3)^2 + (+3)^2} = 4.24 \text{ kN m}$$

2.6.6   MOMENT OF A FORCE ABOUT A GIVEN AXIS

Consider force $\bar{F}$ acting on a rigid body, as shown in Figure 2.24. To find the moment of force $\bar{F}$ about line L, we first determine the moment, $\bar{M}_O$, at point O:

$$\bar{M}_O = \bar{r}_{OA} \times \bar{F} \qquad\qquad (2.35)$$

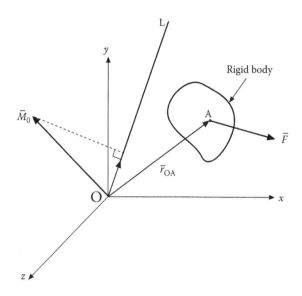

FIGURE 2.24   Moment of a force about a given axis.

Assuming that point O is on line L, projection of $\bar{M}_O$ on line L gives moment M_L of force $\bar{F}$ on line L. Thus, M_L can be determined by the dot product of $\bar{M}_O$ and $\bar{\lambda}$, where $\bar{\lambda}$ is the unit vector on line L. Combinations of dot products of unit vectors i, j, and k are: $i \cdot i = j \cdot j = k \cdot k = 1$ and $i \cdot j = j \cdot k = k \cdot i = 0$.

Example 2.7

Assume that line L passes through points O and B, as shown in Figure 2.25. If the pipe shown in the figure is in the yz plane, determine moment M_L of force $\bar{F} = 100i - 20j - 10k$ N about line L. Point B has coordinates of B (3, 6, 0) m.

SOLUTION

Moment about M_L is given by

$$M_L = \bar{\lambda}_L \cdot \bar{M}_O$$

where unit vector $\bar{\lambda}_L$ on line L is

$$\bar{\lambda}_L = \frac{dx\, i + dy\, j + dz\, k}{d} = \frac{(3 - 0) + (6 - 0) + (0 - 0)}{\sqrt{3^2 + 6^2 + 0^2}}$$

$$\bar{\lambda}_L = \frac{3i + 6j}{\sqrt{45}}$$

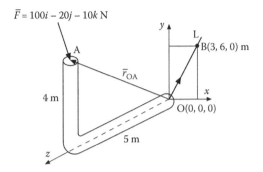

FIGURE 2.25 Pipe loaded by external force.

and moment about point O is

$$\bar{M}_O = \bar{r}_{OA} \times \bar{F}$$

$$\bar{M}_O = (4j + 5k) \times (100i - 20j - 10k)$$

$$\bar{M}_O = 60i + 500j - 400k \text{ N m}$$

Then, moment M_L can be obtained as follows:

$$M_L = \left(\frac{3i + 6j}{\sqrt{45}}\right) \cdot \left(60i + 500j - 400k\right) = \frac{1}{\sqrt{45}}(180 + 3000)$$

$$M_L = 474 \text{ N m}$$

2.6.7 MOMENT OF A COUPLE

Equal, parallel forces acting in opposite directions are called a couple (see Figure 2.26). The perpendicular distance between the action lines of the forces is called the moment arm of the couple. The vector sum of these two forces is zero, but the moment of the couple is not zero. The magnitude of the moment of the couple is given by

$$M = d \times F \tag{2.36}$$

A rigid body shown in Figure 2.26 will rotate with a moment of $d \times F$ about an axis perpendicular to the plane of the couple. This moment magnitude is the same everywhere on the rigid body.

2.7 STRUCTURES IN THREE DIMENSIONS

As in planar trusses, two methods can be used to determine the forces in a space structure: *method of joint* and *method of sections*. In the analysis of space structures, using vector notation can simplify the solution of the problem considerably. For a

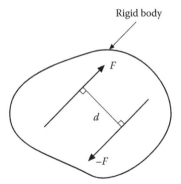

FIGURE 2.26   Rigid body loaded by couple.

given space structure to be in equilibrium, the following two vector equations must be satisfied:

$$\sum \bar{F} = 0$$
$$\sum \bar{M} = 0$$

(2.37)

The equivalent six scalar equations are

$$\sum F_x = 0; \quad \sum F_y = 0; \quad \sum F_z = 0$$
$$\sum M_x = 0; \quad \sum M_y = 0; \quad \sum M_z = 0$$

(2.38)

Table 2.2 can be used when the FBD of a space structure is drawn. Appropriate reaction forces should be used when the structure is isolated from the supports.

Example 2.8

As shown in Figure 2.27, a boom ACD is supported by a ball-and-socket joint at A, by cable DE, and by a sliding bearing at B. If a force of 2 kN is applied at C, determine the force in cable DE and the reaction forces at A and B of the space structure.

SOLUTION

Using Table 2.2, draw an FBD as shown in Figure 2.28. Then, referring to the FBD determine the position vectors $\bar{r}_{AB}$, $\bar{r}_{AC}$, and $\bar{r}_{AD}$ as follows:

$$\bar{r}_{AB} = 3i + 6k$$

$$\bar{r}_{AC} = -3j + 6k$$

$$\bar{r}_{AD} = 3j + 6k$$

TABLE 2.2

Examples of Replacement Reaction Forces to Be Used for Different Types of Supports in a Space Structure

Type of Support	Reaction Forces to Be Replaced
 Slider bearing	
 Thrust bearing	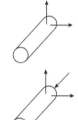
 Fixed support	
 Ball-and-socket joint	
 Contact on rough surface	
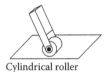 Cylindrical roller	

Tension in cable ED is given by

$$\overline{T}_{DE} = \overline{\lambda}_{DE} \cdot T_{DE}$$

Unit vector $\overline{\lambda}_{DE}$ is

$$\overline{\lambda}_{DE} = \frac{dx\ i + dy\ j + dz\ k}{d} = \frac{4i - 2j - 6k}{\sqrt{4^2 + (-2)^2 + (-6)^2}}$$

$$\overline{\lambda}_{DE} = 0.53i - 0.27j - 0.802k$$

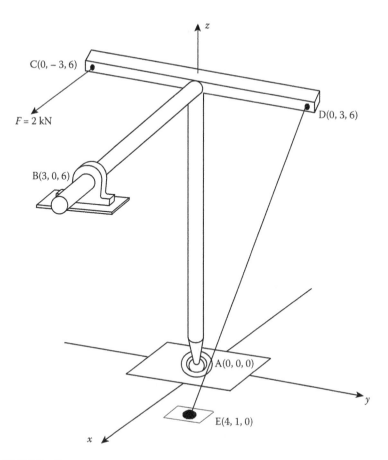

FIGURE 2.27 Space structure.

Then the tension in DE can be calculated as

$$\bar{T}_{DE} = 0.53T_{DE}i - 0.27T_{DE}j - 0.802T_{DE}k$$

Using the FBD shown in Figure 2.28, determine the reaction forces at support B. Applying the moment equilibrium equation at point A gives

$$\sum \bar{M}_A = 0$$

$$\sum \bar{M}_A = \bar{r}_{AC} \times \bar{F} + \bar{r}_{AD} \times \bar{T} + \bar{r}_{AB} \times \bar{B} = 0$$

Each term of the moment equation given above can be determined as follows:

$$\bar{r}_{AC} \times \bar{F} = (-3j + 6k) \times 2i = 6k + 12j$$

$$\bar{r}_{AD} \times \bar{T} = (3j + 6k) \times (0.53T_{DE}i - 0.27T_{DE}j - 0.802T_{DE}k)$$

$$= -0.78T_{DE}i + 3.18T_{DE}j - 1.59T_{DE}k$$

$$\bar{r}_{AB} \times \bar{B} = (3i + 6k) \times (B_y j + B_z k) = -6B_y i - 3B_z j + 3B_y k$$

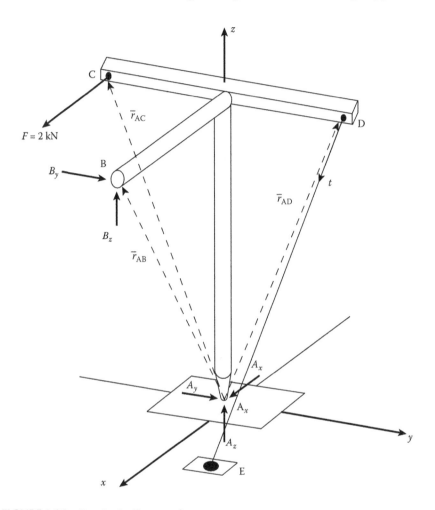

FIGURE 2.28 Free-body diagram of space structure.

Substituting each term into the moment equation yields

$$\sum \bar{M}_A = (6k + 12j) + (-0.78T_{DE}i + 3.18T_{DE}j - 1.59T_{DE}k)$$
$$+ (-6B_yi - 3B_zj + 3B_yk) = 0$$

Rearranging the above equation yields

$$\sum \bar{M}_A = (-0.78T_{DE} - 6B_y)i + (12 + 3.18T_{DE} - 3B_z)j$$
$$+ (6 - 1.59T_{DE} + 3B_y)k = 0$$

Since each term of the above equation must equal zero, the following three equations can be obtained:

$$-0.78T_{DE} - 6B_y = 0 \tag{a}$$

$$12 + 3.18T_{DE} - 3B_z = 0 \tag{b}$$

$$6 - 1.59T_{DE} + 3B_y = 0 \tag{c}$$

From Equation (a), we obtain

$$B_y = -\frac{0.78}{6}T_{DE}$$

Substituting B_y into Equation (c), the magnitude of tension in DE, $T_{DE} = 3.03$ kN, can be obtained. Hence, the tension force in DE is

$$\overline{T}_{DE} = -0.78T_{DE}i + 3.18T_{DE}j - 1.59T_{DE}k$$

$$= (-0.78 \times 3.03i + 3.18 \times 3.03j - 1.59 \times 3.03k)$$

$$\overline{T}_{DE} = -2.36i + 9.64j - 4.82k \text{ kN}$$

From Equation (a)

$$B_y = -\frac{0.78}{6}T_{DE} = -\frac{0.78}{6}(3.03) = -0.40 \text{ kN}$$

The negative sign of B_y indicates that the direction of the force on the FBD should be in the opposite direction. Note that a positive sign would be obtained if the direction had been assumed correctly in the negative Y direction. From Equation (b), the value of B_z can be determined as follows:

$$B_z = \frac{12 + 3.18T_{DE}}{3} = \frac{12 + 3.18(3.03)}{3} = 7.21 \text{ kN}$$

Then, the reaction force at sliding bearing B is

$$\overline{B} = -0.40j + 7.21k \text{ kN}$$

Finally, applying equilibrium equation $\sum \overline{F} = 0$ to determine the reaction forces at point A yields:

$$\sum \overline{F} = (A_xi + A_yj + A_zk) + (2i) + (-0.40j + 7.21k)$$

$$+ (-2.36i + 9.64j - 4.82k) = 0$$

or

$$\overline{A} = (A_xi + A_yj + A_zk) = 0.36i - 9.24j - 2.39k$$

Therefore, the reaction force components at the ball-and-socket joint are

$$A_x = 0.36 \text{ kN}$$
$$A_y = -9.24 \text{ kN}$$
$$A_z = -2.39 \text{ kN}$$

2.8 TRUSSES

A truss is a structure consisting of a number of members fastened together at their ends in such a manner as to form a rigid body. A truss may be used to support a larger load or span a greater distance than can be done by a single member. Trusses are often used in bridge and roof construction, and so on.

When three members are connected by a frictionless pin at their ends, as shown in Figure 2.29, they form a rigid structure. The three members forming a triangle is a basic truss element. When truss members all lie in one plane they are called plane trusses. If truss members do not all lie in one plane they are called space trusses.

If four members are connected by frictionless pins, as shown Figure 2.30, the resultant structure is not rigid and when force is applied the structure can collapse. These types of structures are called mechanisms and they are the basis of all

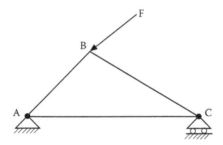

FIGURE 2.29   Basic truss structure with three members.

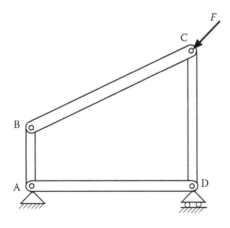

FIGURE 2.30   Truss structure with four members.

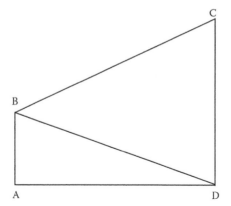

FIGURE 2.31 Rigid truss structure.

FIGURE 2.32 Two-force member.

machines. For example, as shown in Figure 2.30, when AB is in motion, it produces a controlled motion of BC and CD.

If an additional member from B to D is added to the four-link mechanism, as shown in Figure 2.31, it will restore rigidity. The addition of the member BD changes the truss structure into two triangles, which satisfies the rule of rigidity mentioned earlier.

Trusses are designed so that applied loads act at the end of the truss members. In general, it is assumed that the weight of truss members is negligible in comparison with the applied load. Since truss members are two-force members, the line of action of the forces exerted by the pins is along the truss member (see Figure 2.32, FBD of a truss member).

Members that are stretched, as shown in Figure 2.32, are said to be in tension, T, and those that are shortened are said to be in compression, C. The following two common methods are used for truss analysis.

2.8.1 METHOD OF JOINT

To calculate the forces acting on the joints, the following steps can be used. When using this method to calculate forces in truss members, equilibrium equations are applied to each joint.

1. Assume all members of a truss are two-force members.
2. Draw an FBD of the entire truss structure and calculate the reaction forces at the supports.

3. Using the reaction forces, determine the forces at the starting joint that has two unknowns. Draw an FBD of the joint assuming that all members attached to this joint are in tension. If the result yields a negative value, then the members are under compression.
4. Determine all the unknown forces at each consecutive joint.

Example 2.9

Determine the forces in members AB, AH, BH, and BC of the truss shown in Figure 2.33 by using the method of joint analysis.

SOLUTION

To determine the unknown reaction forces, draw an FBD of the entire structure as shown in Figure 2.34, and apply equilibrium equations as follows:

$$\sum F_x = 0 \Rightarrow A_x = 0$$

$$\sum F_y = 0$$

$$A_y + E_y - 4000 - 2000 = 0$$

$$A_y + E_y = 6000$$

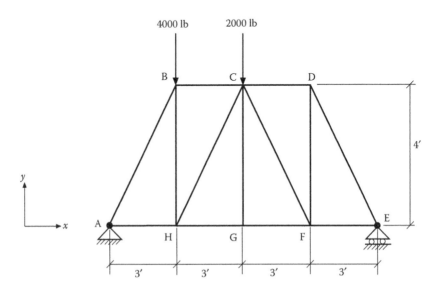

FIGURE 2.33 Truss structure.

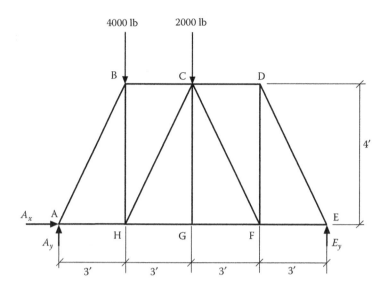

FIGURE 2.34 Free-body diagram of truss structure.

Applying the moment equation for equilibrium yields
$\sum M_A = 0$ (assuming counterclockwise moments are positive),

$$-4000 \times 3 - 2000 \times 6 + E_y \times 12 = 0$$

$$E_y = \frac{12{,}000 + 12{,}000}{12} = 2000 \text{ lb}$$

As shown in Figure 2.35, draw an FBD of joints A and B and apply equilibrium equations. Note that internal forces in all members are assumed to be in tension, as shown in Figure 2.35.

Equilibrium equations at joint A are

$$\sum F_y = 0 + \uparrow$$

$$4000 + \frac{1}{\sqrt{2}} T_{AB} = 0$$

$$T_{AB} = -5657 \text{ lb}$$

where the negative value of T_{AB} indicates that member AB is under compression; and

$$\sum F_x = 0 + \rightarrow$$

$$A_x + \frac{1}{\sqrt{2}} T_{AB} + T_{AH} = 0$$

$$T_{AH} = A_x - \frac{1}{\sqrt{2}} T_{AB} = 0 + \left[-\frac{1}{\sqrt{2}} \times (-5657) \right] = 4000 \text{ lb}$$

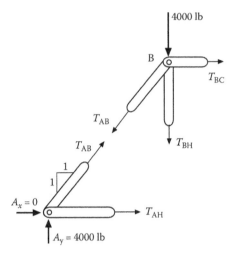

FIGURE 2.35 Free-body diagram of joints A and B.

where the positive value of T_{AH} indicates that member AH is under tension.

To determine the forces in members T_{BC} and T_{BH}, apply the equilibrium equations at joint B. From the previous calculation, $T_{AB} = -5657$ lb.

$$\sum F_x = 0 + \rightarrow$$

$$T_{BC} - \frac{1}{\sqrt{2}} T_{AB} = 0$$

$$T_{BC} = \frac{1}{\sqrt{2}}(-5657) = -4000 \text{ lb (compression)}$$

$$\sum F_y = 0 + \uparrow$$

$$-4000 - T_{BH} - \frac{1}{\sqrt{2}} T_{AB} = 0$$

$$T_{BH} = -4000 - \frac{1}{\sqrt{2}}(-5657) = 0$$

Since the resulting force value of member T_{BH} is zero, it is called a *zero-force member*. Note that zero-force members under one loading condition may not be zero-force members under a different loading condition.

2.8.2 METHOD OF SECTION

In this method, forces on members are determined by imagining that the truss is cut into sections and applying equilibrium equations to each of these sections. To illustrate this method, examine Figure 2.35. To determine forces in members CH and HG, for example, the truss is cut into sections along a line a–a. Note that line a–a cuts the interested members in question (see Figure 2.36). Again, it is convenient to assume that all members are in tension. FBDs of two sections of the truss are shown in Figure 2.36.

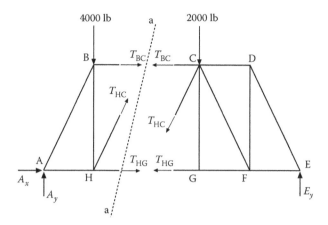

FIGURE 2.36 Method of section.

Any portion of the truss can be used to determine forces in members T_{BC}, T_{HC}, and T_{HG}. Applying equilibrium equations to the left portion of the truss yields

$$\sum F_y = 0 + \uparrow$$
$$A_y - 4000 + T_{HC} \sin 45° = 0$$
$$T_{HC} = \frac{4000 - 4000}{\sin 45°} = 0$$

Applying the moment equation at point A gives
$\sum M_H = 0$ (assuming counterclockwise moments are positive)

$$-4000 \times 3 - T_{BC} \times 3 = 0$$
$$T_{BC} = -4000 \text{ lb}$$

Forces for other members of the truss structure can be determined using the same procedures.

2.9 MACHINES AND FRAMES

So far, the force analysis used has been relatively simple. Since the system consisted of only one major element, one FBD with one set of equilibrium equations was used in the analysis. Machines and frames consist of various members that together form complex systems. Force analyses of such systems and their individual elements require no new principles but the use of a large number of FBDs and equilibrium equations. In general, these structures are loaded with two or more forces.

Machines are complex systems that contain moving members. In general, they are designed to convert an input force to an output force. Frames may have a large number of members and are designed to carry heavy loads in a fixed position. Force analyses of machines and frames may require using two- and three-force members.

Example 2.10

Determine the gripping force P developed in the vise grip pliers when 500 lb force is applied to the handles, as shown in Figure 2.37.

SOLUTION

Applying equilibrium equations to the entire vise grip pliers gives no information. Thus, consider FBDs of the individual members as shown in Figure 2.38. First, consider member AB. Since this member is a two-force member, the line of action of the force must pass through points A and B. Then the direction of the force at B and A is

$$\tan\alpha^{-1} = \frac{0.035}{0.075} = 25°$$

From Figure 2.38b, the components of the force at B are

$$B_x = B \times \cos 25° = 0.906B$$
$$B_y = B \times \sin 25° = 0.423B$$

The same force components act at point A (keeping in mind the two-force member concept):

$$A_x = 0.906B$$
$$A_y = 0.423B$$

Consider now the FBD in Figure 2.38c. Force components B_x and B_y will be the same for member AB with opposite directions. Taking the sum of the moments about point C yields

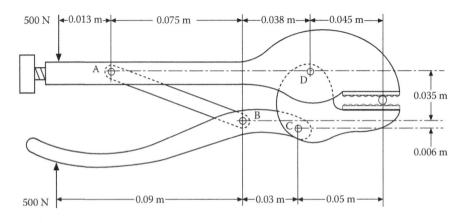

FIGURE 2.37   Vise grip plier.

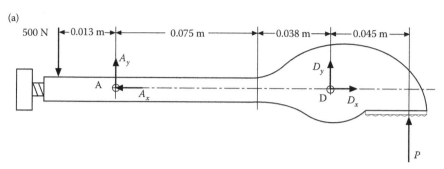

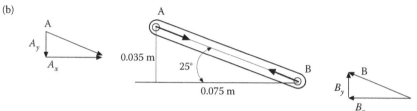

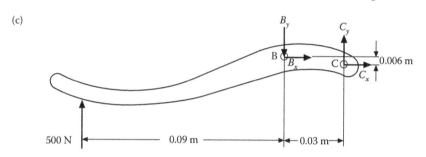

FIGURE 2.38 Free-body diagrams of vise grip plier.

$\sum M_C = 0$ (assuming counterclockwise moments are positive)

$$-0.12 \times 500 - 0.006 B_x + 0.03 B_y$$

or

$$-0.12 \times 500 - 0.006(0.906B) + 0.03(0.423B) = 0$$

From the above equation, calculating the unknown yields

$$B = 8271 \text{ N}$$

Taking the sum of the moments about point D, as shown in Figure 2.38a, gives.

$$0.045P - 0.113(3499) + 0.126 \times 500 = 0$$
$$0.045P - 0.113 \, A_y + 0.126 \times 500 = 0$$

But

$$A_y = B_y = 0.423 \times B = 0.423 \times 8271 = 3499 \text{ N}$$

and substituting value of A_y solving for P, we obtain

$$P = 7386 \text{ N}$$

Example 2.11

Determine the components of the forces exerted by the axle of the pulley carrying a load with a mass of 459.2 kg on all the joints, as shown in Figure 2.39. Assume that the weight of the structure is negligible.

SOLUTION

To determine the reaction forces at supports A and F, consider the FBD of the pulley system shown in Figure 2.40.

Taking the sum of moments about point A yields,

$\sum M_A = 0$ (assuming counterclockwise moments are positive)

$$2.2F_y - 4.7 \times 4500$$

$$F_y = 9614 \text{ N}$$

Summation of forces in the y direction gives

$$\sum F_y = 0 + \uparrow$$

$$A_y + F_y - 4500 = 0$$

$$A_y + 9614 - 4500 = 0$$

$$A_y = -5114 \text{ N}$$

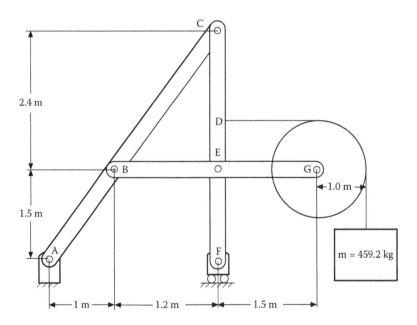

FIGURE 2.39 Pulley system.

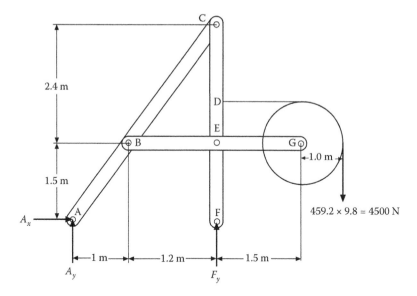

FIGURE 2.40 Free-body diagram of pulley system.

The negative sign indicates that the direction of A_y should be downward. Using the FBD in Figure 2.41a and applying equilibrium equations gives the forces at point G:

$$\sum F_x = 0 + \rightarrow$$

$$G_x - 4500 = 0 \Rightarrow G_x = 4500 \text{ lb}$$

$$\sum F_y = 0 + \uparrow$$

$$G_y - 4500 = 0 \Rightarrow G_y = 4500 \text{ lb}$$

Since the solution yields positive values for G_x and G_y, the assumed force directions are correct.

Similarly, determine the unknown forces at joints B and E by using the FBD in Figure 2.41b. Note that point G is a common point for both the pulley and member BG. Thus, use the same forces with opposite direction at point G on member BG. Summation of moments about point B yields

$$\sum M_B = 0$$

$$1.2E_y - 2.7G_y = 0$$

$$E_y = \frac{2.7 \times 4500}{1.2} = 10125 \text{ N}$$

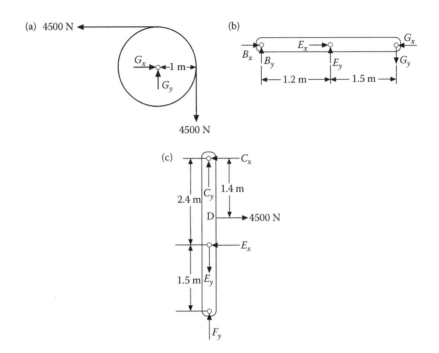

FIGURE 2.41 Free-body diagrams of pulley system members.

Taking moment about point E

$$\sum M_E = 0$$

$$1.2B_y - 1.5G_y = 0$$

$$B_y = \frac{1.5 \times 4500}{1.2} = 5625 \text{ N}$$

$$\sum F_x = 0 + \rightarrow$$

$$B_x + E_x - G_x = 0$$

$$B_x + E_x = 4500$$

Consider now the FBD of member CF

$$\sum M_C = 0$$

$$1.5 \times 4500 - 2.4E_x = 0$$

$$E_x = \frac{1.5 \times 4500}{2.4} = 2812.5 \text{ N}$$

$$\sum F_x = 0 + \rightarrow$$

$$-C_x - E_x + 4500 = 0$$

$$C_x = 4500 - 2812.5 = 1687.5 \text{ N}$$

$$\sum F_y = 0 + \uparrow$$

$$C_y - E_y + F_y = 0$$

$$C_y = 10125 - 9614 = 511 \, \text{N}$$

and

$$B_x + E_x = 4500$$

$$B_x = 4500 - 2812.5 = 1687.5 \, \text{N}$$

2.10 FRICTION

When a body slides or tends to slide relative to another body, a friction force is created between the surfaces in contact. This friction force is tangent to the plane of contact and resists the motion of the body upon which it acts.

Machines, holding and fastening devices, brakes, walking traction, and vehicle movement all require frictional forces in order to function. However, friction causes loss of power and wears, which are undesirable. Consider a block sitting on a rough surface being subjected to horizontal force P, as shown in Figure 2.42a. When applied force P is zero, the block is in equilibrium and friction force F is also zero (see Figure 2.42b). When P increases the friction force increases correspondingly to maintain equilibrium until it reaches its maximum value (unstable equilibrium). Any further increase in P results in motion, and friction force F then becomes a function of the velocity of the block.

As shown in Figure 2.43, friction force F does not stay at its maximum value, but decreases rapidly when motion starts, reaching a fairly constant kinetic friction value for a corresponding constant velocity of the block.

2.10.1 COEFFICIENT OF FRICTION

The static coefficient of friction, μ_s, is defined as the ratio of maximum static frictional force, F_{max}, to the normal force, N. The static coefficient of friction is given by

$$\mu_s = \frac{F_{max}}{N} \tag{2.39}$$

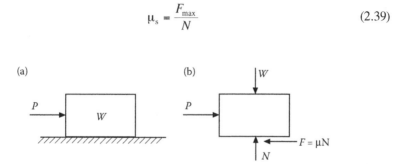

FIGURE 2.42 (a) Block sitting on a rough surface and (b) free-body diagram.

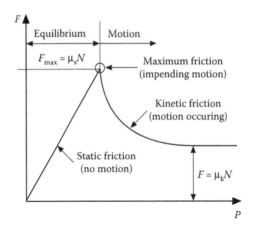

FIGURE 2.43   Static and kinetic friction coefficients.

The kinetic coefficient of friction in motion is given by

$$\mu_k = \frac{F}{N} \qquad\qquad (2.40)$$

2.10.2 ANGLES OF STATIC AND KINETIC FRICTION

As shown in Figure 2.44, angle φ_s between R and N is defined as the angle of friction. For impending motion, φ_s is given by

$$\varphi_s = \tan^{-1} \mu_s \qquad\qquad (2.41)$$

and, when there is relative motion between the two surface, φ_k is given by

$$\varphi_k = \tan^{-1} \mu_k \qquad\qquad (2.42)$$

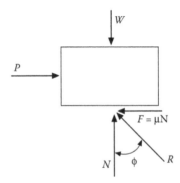

FIGURE 2.44   Angle of friction.

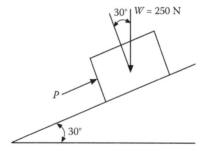

FIGURE 2.45 Inclined planes.

Example 2.12

A metal block with a weight of 250 N is in contact with a plane inclined at 30° to the horizontal axis, as shown in Figure 2.45. Determine force P required to

   a. Prevent the block from sliding down the incline ($\mu_s = 0.25$).
   b. Start it moving up the incline.

SOLUTION

   a. Figure 2.46a shows the FBD of the block when motion down the inclined plane is impending.
     Applying equilibrium equations yields

$$\sum F_y = 0$$

$$N - 250 \times \cos 30° = 0$$

$$N = 216.5 \text{ N}$$

$$\sum F_x = 0$$

$$P + \mu N - 250 \times \sin 30° = 0$$

$$P = 125 - 0.25 \times 216.5 = 70.88 \text{ N}$$

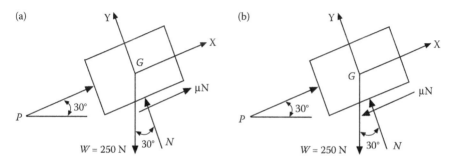

FIGURE 2.46 Free-body diagram (a) when motion is down and (b) when motion is up.

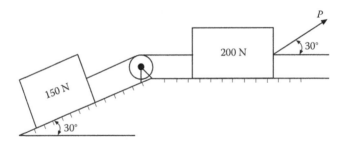

FIGURE 2.47   Impending motion.

b. Figure 2.46b shows the FBD of the block when motion up the inclined plane is impending. In this case, friction force F acts down the inclined plane to resist motion. Again, applying equilibrium equations gives

$$\sum F_x = 0$$

$$P - \mu N - 250 \times \sin 30° = 0$$

$$P = 0.25 \times 216.5 + 125 = 179.13 \text{ N}$$

Example 2.13

Determine the minimum value of P that will just start the system of blocks shown in Figure 2.47 moving to the right. Assume that the coefficient of friction under each block is 0.4.

SOLUTION

Consider the FBDs of each block as shown in Figures 2.48a and b. Applying equilibrium equations to the FBD in Figure 2.48a yields

$$\sum F_y = 0$$

$$N_1 - 150 \times \cos 30° = 0$$

$$N_1 = 130 \text{ N}$$

and

$$\sum F_x = 0$$

$$T - \mu N_1 - 150 \times \sin 30° = 0$$

$$T = 0.4 \times 130 + 75 = 127 \text{ N}$$

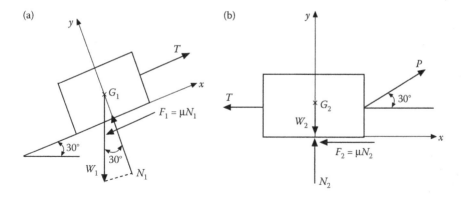

FIGURE 2.48 Free-body diagram of weight (a) 150 N and (b) 200 N.

Similarly, applying equilibrium equations to the FBD in Figure 2.48b yields the minimum value of P that will just start the system of blocks:

$$\Sigma F_y = 0$$

$$-W_2 + N_2 + P \sin 30° = 0$$

$$-200 + N_2 + 0.5P = 0$$

$$N_2 = 200 - 0.5P$$

$$\Sigma F_x = 0$$

$$-T + P \times \cos 30° - 0.4N_2 = 0$$

$$-127 + 0.866P - 0.4(200 - 0.5P) = 0$$

$$P = 194.2 \text{ N}$$

2.11 PROPERTIES OF PLANE AREAS

In studying the strength of materials, it is necessary to find the centroid of an area in order to determine the location of the neutral axis in the bending of beams, for example. For bending, it is known that stress along the neutral axis is zero. An axis passing through the centroid of an area is known as centroidal axis. If the body is homogeneous (i.e., with uniform mass distribution), the centroid and mass center coincide.

Centroid: The equations to determine the coordinates of the centroid of an area are defined by the following integrals:

$$\bar{x} = \frac{\int_A x\,dA}{\int_A dA} \quad \text{and} \quad \bar{y} = \frac{\int_A y\,dA}{\int_A dA} \tag{2.43}$$

where dA is the differential element of A, located at any arbitrary x and y, as shown in Figure 2.49. Integrals $\equiv x$ dA or $\equiv y$ dA are often referred to as the first moment of area with respect to the x- and y-axis, respectively.

To determine the centroid by integration, the area is divided into differential elements so that

1. All points of the differential elements are located at the same distance from the axis of moments under consideration.
2. If the position of the centroid of a differential element is known, the moment of the differential element about the axis under consideration is the product of the differential elements area and the distance of its centroid from the axis.

The moment of inertia (second moment of area), I_x and I_y, of an area with respect to the x- and y-axis, respectively, is given by

$$I_x = \int_A x^2 \, dA \quad \text{and} \quad I_y = \int_A y^2 \, dA \tag{2.44}$$

The second moment of inertia is also called the rectangular moment of inertia. The second moment of the element of area shown in Figure 2.49 with respect to an axis through point O perpendicular to the plane of the paper is called the polar moment of inertia and is given by

$$J_o = \int_A r^2 \, dA = \int_A (x^2 + y^2) \, dA = I_x + I_y \tag{2.45}$$

From the above equation, it is clear that the polar moment of inertia is equal to the sum of the rectangular moment of inertia with respect to any two perpendicular axes intersecting the polar axis.

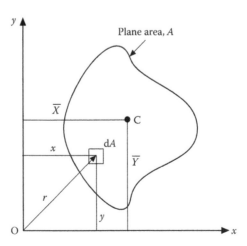

FIGURE 2.49 Centroid of an area.

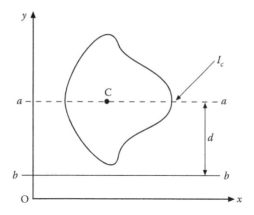

FIGURE 2.50 Inertia about the parallel axis.

2.11.1 Parallel Axis Theorem for Areas

Once the moment of inertia of an area has been determined with respect to one axis, it is often necessary to determine the moment of inertia with respect to a parallel axis. For example, if the moment of inertia, I_c, about the centroidal axis is known (see Figure 2.50), the moment of inertia about the axis b–b parallel to axis a–a is given by

$$I_b = I_c + Ad^2 \tag{2.46}$$

where d is the distance between a–a axis and the parallel axis b–b. It can be also shown that

$$J_b = J_c + Ad^2 \tag{2.47}$$

From the above equations, it is evident that the moment of inertia of any area with respect to an axis through its centroid is less than that for any parallel axis.

2.11.2 Radius of Gyration of Area

Although the *radius of gyration of area* has no physical significance, it is common practice to introduce it in some structural engineering applications. If the moment of inertia is known, the radius of gyration k about x, y, and point O is defined as

$$k_x = \sqrt{\frac{I_x}{A}}, \quad k_y = \sqrt{\frac{I_y}{A}}, \quad k_O = \sqrt{\frac{J_O}{A}} \tag{2.48}$$

The radii of gyration are related as follows:

$$k_o^2 = k_x^2 + k_y^2 \tag{2.49}$$

2.11.3 Moment of Inertia of Composite Areas

In general, a composite area is composed of combinations of geometrical shapes such as rectangles, triangles, and circles. If a given composite area can be divided into known geometrical shapes, the moment of inertia of that area with respect to any axis is equal to the sum of moments of inertia of the geometrical shapes with respect to the same axis. If an area is removed from any given larger area, its moment of inertia is subtracted from the moment of inertia of the larger area to find the net moment of inertia. Table A1.11 (see Appendix 1) shows the centroids and mass moment of inertia of common geometric shapes.

Example 2.14

Determine the moment of inertia about the x-axis and the centroidal axis of the rectangular shape shown in Figure 2.51.

SOLUTION

The moment of inertia about the x-axis is given by

$$I_x = \int y^2 \, dA$$

where

$$dA = b \, dy$$

Substituting the above yields

$$I_x = b \int y^2 \, dy = b \left. \frac{y^3}{3} \right|_0^h = \frac{bh^3}{3}$$

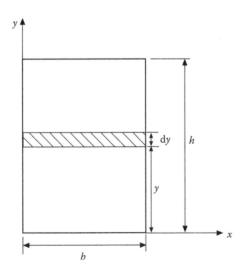

FIGURE 2.51 Moment of inertia of a rectangular shape.

The moment of inertia about the centroidal axis can be determined from the parallel axis theorem $I_x = I_c + Ad^2$. Then the moment of inertia about the centroidal axis, I_c, is

$$I_c = I_x - Ad^2$$

$$I_c = \frac{bh^3}{3} - (bh)\left(\frac{h}{2}\right)^2 = \frac{bh^3}{12}$$

Table A1.5 lists the moment of inertia of some common areas about their centroidal axis (see Appendix 1).

Example 2.15

Determine the centroid of the shaded area shown in Figure 2.52. Assume the equation of the curve given in the figure is $y = x^2$.

SOLUTION

Taking a vertical strip parallel to the y-axis satisfies the requirement that all points in this element are the same distance from the y-axis. The area of the shaded strip is

$$dA = y \, dx$$

$$A \int y \, dx = \int_0^4 x^2 \, dx = \left.\frac{x^3}{3}\right|_0^4 = 21.33 \text{ mm}^2$$

$$\bar{x} = \frac{\int x \, dA}{\int dA} = \frac{\int xy \, dx}{21.33} = \frac{\int_0^4 xx^2 \, dx}{21.33} = \frac{1}{21.33}\left.\frac{x^4}{4}\right|_0^4 = 3 \text{ mm}$$

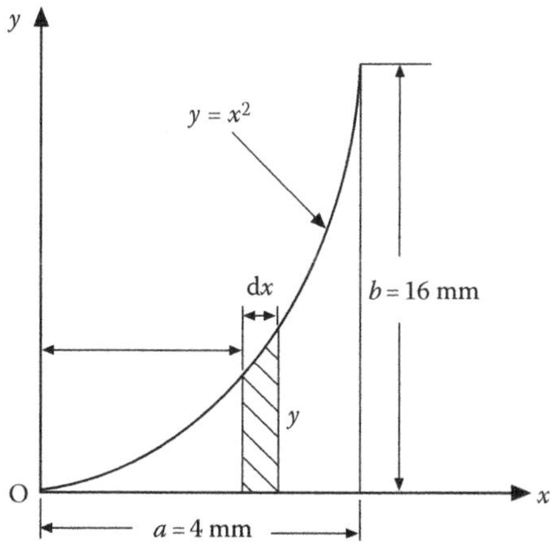

FIGURE 2.52 Moment of inertia of a rectangular shape (using vertical strip).

To determine $\bar{y}$, we can use a horizontal strip or we can use the same vertical strip. If the vertical strip is used, each point of the element is not the same distance from the x-axis. The moment of the area about the x-axis, $y\,dA$, is equal to the product of its centroidal coordinate, $1/2y$, multiplied by the area, dA. Thus

$$\bar{y} = \int y\,dA = \int \frac{1}{2}y\,dA = \int \frac{1}{2}y \cdot y\,dx = \frac{1}{2}\int y^2\,dx$$

Therefore

$$\bar{y} = \frac{\int y\,dA}{\int dA} = \frac{\frac{1}{2}\int y^2\,dA}{\int dA} = \frac{\frac{1}{2}\int y^2\,dA}{21.33} = \frac{\int y^2\,dA}{42.66} = \frac{1}{42.66}\int (x^2)^2\,dx$$

$$\bar{y} = \frac{1}{42.66}\left.\frac{x^5}{5}\right|_0^4 = 4.8 \text{ mm}$$

The same problem can be solved by using a horizontal strip, $dA = x\,dy$, as shown in Figure 2.53. The area of the shaded strip is

$$dA = (4 - x)\,dy$$

where $x = \sqrt{y}$

$$\bar{y} = \frac{\int y\,dA}{\int dA} = \frac{\int y(4 - x)\,dy}{21.33} = \frac{\int y(4 - \sqrt{y})\,dy}{21.33} = \frac{\int \left(4y - y^{3/2}\right)\,dy}{21.33}$$

$$\bar{y} = \frac{1}{21.33}\left.\left|4\frac{y^2}{2} - \frac{y^{5/2}}{5/2}\right|\right._0^4 = 4.8 \text{ mm}$$

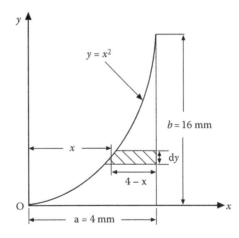

FIGURE 2.53 Moment of inertia of a rectangular shape (using horizontal strip).

Example 2.16

Determine the moment of inertia of the T shape, shown in Figure 2.54, about the horizontal axis passing through the centroid.

SOLUTION

First locate the centroid of the composite area:

$$\bar{y} = \frac{\int y \, dA}{\int dA} = \frac{\bar{y}_1 A_1 + \bar{y}_2 A_2}{A_1 + A_2} = \frac{3.5(1 \times 4) + 1.5(3 \times 1)}{(1 \times 4) + (3 \times 1)} = 2.64 \text{ in.}$$

The moment of inertia of the first area about its centroid is

$$I_{c_1}^1 = \frac{bh^3}{12} = \frac{4 \times 1^3}{12} = \frac{1}{3} \text{ in.}^4$$

The moment of inertia of the second area about its centroid is

$$I_{c_2}^2 = \frac{bh^3}{12} = \frac{1 \times 3^3}{12} = \frac{3}{4} \text{ in.}^4$$

Using the parallel axis theorem

$$I_c^1 = I_{c_1}^1 + A_1 d_1^2 = \frac{1}{3} + (1 \times 4)(3.5 - 2.64)^2 = 3.29 \text{ in.}^4$$

$$I_c^2 = I_{c_2}^2 + A_2 d_2^2 = \frac{3}{4} + (1 \times 3)(2.64 - 1.5)^2 = 4.65 \text{ in.}^4$$

Then the moment of inertia of the entire composite shape is

$$I_c = I_c^1 + I_c^2 = 3.29 + 4.65 = 7.94 \text{ in.}^4$$

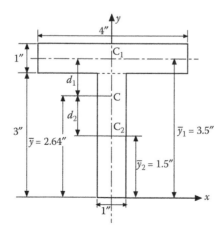

FIGURE 2.54 Moment of inertia of a T shape.

PROBLEMS

2.1. Using the following figure, determine the magnitude of F so that the resultant force will be vertical.

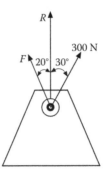

2.2. Using the following figure, determine the reactions at points A and B on the wall. Assume the wall is vertical.

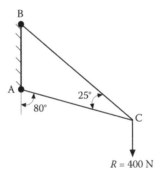

2.3. Using the following figure, determine the angle required for a resultant force magnitude of 250 lb.

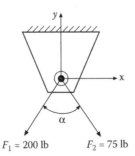

2.4. Determine the resultant force in terms of unit vectors i and j applied on a pin connection shown in the following figure.

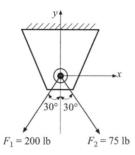

2.5. Using the following figure, determine the components of vertical force in the direction of support cable AB and truss member BC.

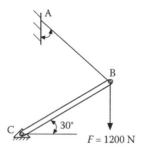

2.6. As shown in the following figure, cables AB and BC exert forces $F_1 = 150$ lb and $F_2 = 200$ lb on the block. Determine (a) the projection of F_2 onto F_1, and (b) the projection of F_1 onto F_2.

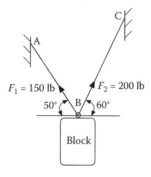

2.7. The coplanar force system shown in the following figure has a 260-lb resultant force in the positive direction of *y*. Determine the magnitude and direction O of force *P*.

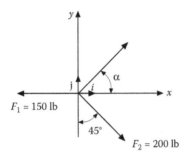

2.8. For the cable system given in the following figure, determine the value of α for equilibrium.

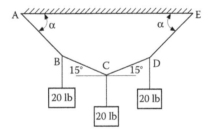

2.9. As shown in the following figure, a 500-lb weight is supported by two cables AC and BC. Determine the value of α for minimum tension in AC and find the corresponding values of tension in cables AC and BC.

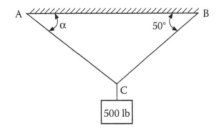

2.10. Determine force F required to hold a weight of 300 lb in equilibrium, as shown in the following figure. Ignore the friction effect.

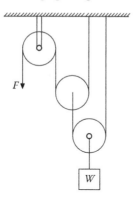

2.11. Determine the moment of force about line OA if the pipe is in the yz plane, as shown in the following figure.

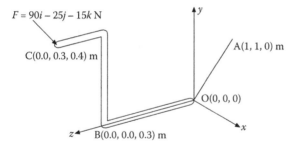

2.12. Using the following figure, if the applied force at point A is $\overline{F} = 50i + 100j + 200k$ N, determine the moment of $\overline{F}$ about line BD.

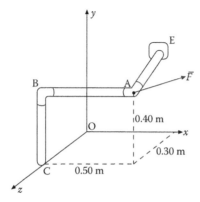

2.13. Determine the moment of $\bar{T}$ = 5000 lb about line AC in the following figure.

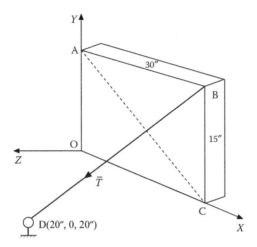

2.14. As shown in the following figure, a boom ACD is supported by a ball-and-socket joint at A, by cable DE, and by a sliding bearing at B. If the force of 2 kN is applied at C, determine the force in cable DE and the reaction forces at A and B of the space structure.

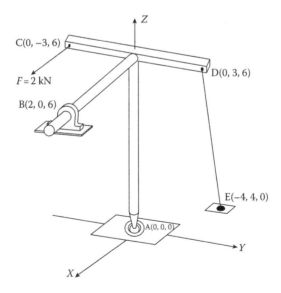

2.15. As shown in the following figure, a vertical mast AD is supported by a ball-and-socket joint at A by cables DB and DC. If the force

$\overline{F} = 100\overline{i} + 100\overline{j}$ N is applied at E, determine the reaction forces at A and forces in cables DB and DC.

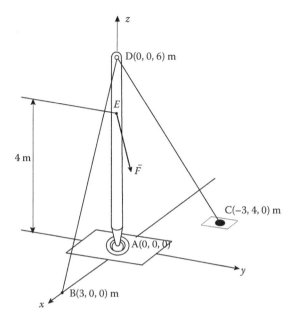

2.16. Determine the reaction forces and the forces in members BD and BE of the truss shown in the following figure.

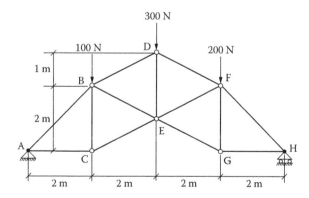

2.17. Using the following figure, locate the centroid and determine the moment of inertia of the rectangular area about the x-axis passing through the centroid.

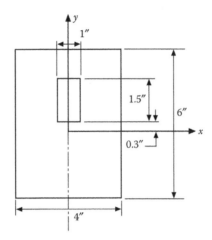

2.18. Determine the moment of inertia and radius of gyration of the area given in the following figure about the x-axis passing through the centroid.

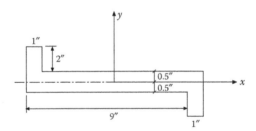

BIBLIOGRAPHY

Bedford, A. and Fowler, W. *Engineering Mechanics: Statics*, Addison-Wesley, Reading, MA, 1998.

Hibbeler, R. C. *Statics and Mechanics of Materials*, Prentice-Hall, Englewood Cliffs, NJ, 1992.

Higdon, A., Stiles, W. B., Davis, A. W., Evces, C. R., and Weese, J. A. *Engineering Mechanics, Volume I: Statics*, Prentice-Hall, Englewood Cliffs, NJ, 1979.

Singer, F. L. *Engineering Mechanics, Part I: Statics*, Harper & Row, New York, 1975.

3 Dynamics

3.1 INTRODUCTION

Dynamics deals with the motion of bodies under the action of forces. It is studied under two distinct parts: kinematics, which deals with the motion of a body without consideration of the forces causing the motion, and kinetics, which is the study of body motion taking into consideration the forces causing it. In this chapter, kinematics and kinetics of a rigid body will be briefly discussed.

3.2 KINEMATICS OF A RIGID BODY

In dynamics, the resultant force–couple system is not zero and this causes a change in the state of motion of the rigid body on which it acts. In general, the resultant force–couple system is applied at the center of mass of a rigid body. Types of rigid-body motion can be classified into three categories: translation, fixed-axis rotation, and general plane motion.

3.2.1 TRANSLATION

If the resultant force system consists of only a single force passing through the center of mass of a rigid body, the body will move in the direction of the resultant force R. Translation is a motion without rotation such that any line on the rigid body moves parallel to itself. There are two types of translational motion: (a) rectilinear, where the line in motion moves along parallel straight lines as shown in Figure 3.1, and (b) curvilinear, where the line in motion moves along parallel curved lines as shown in Figure 3.2.

3.2.1.1 Velocity in Translational Motion

In translational motion, all points on the body have the same velocity V.

$$\bar{V}_{\mathrm{B}} = \bar{V}_{\mathrm{A}} \tag{3.1}$$

Velocity is defined as the time rate of change of position and can be determined as

$$V = \frac{\mathrm{d}s}{\mathrm{d}t} \tag{3.2}$$

where $\mathrm{d}s$ is the change in position and $\mathrm{d}t$ is the elapsed time between two points.

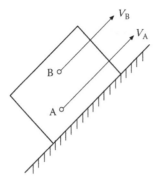

FIGURE 3.1   Rectilinear translation.

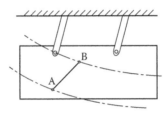

FIGURE 3.2   Curvilinear translation.

3.2.1.2 Acceleration in Translational Motion

In translational motion, all points on the body have acceleration $\vec{a}$.

$$\bar{a}_A = \bar{a}_B \qquad (3.3)$$

Acceleration is defined as the time rate of change of velocity and can be determined as

$$a = \frac{dV}{dt} \qquad (3.4)$$

where dV is the change in velocity and dt is the elapsed time between two points.

If the rigid body is in translational motion, all points on the body have the same velocity and acceleration as the center of mass. From Equations 3.2 and 3.4, eliminating dt we obtain

$$a\,ds = V\,dV \qquad (3.5)$$

3.2.2 FIXED-AXIS ROTATION

If the resultant force system is a couple M as shown in Figure 3.3, the body will move such that all the points on a straight line will have zero velocity relative to an axis passing through its center of mass, and the body will spin relative to this reference axis.

3.2.2.1 Angular Components of Fixed-Axis Rotation

If θ is the angular position of point P measured from the fixed line, $d\theta$ is the angular displacement of point P and the direction of $d\theta$ is found by the right-hand rule (see Figure 3.3).

Angular velocity ω is the time rate change of the angular position and is given as

$$\omega = \frac{d\theta}{dt} \tag{3.6}$$

where ω has the same direction as $d\theta$.

Angular acceleration α is the time rate change of the angular velocity and is given as

$$\alpha = \frac{d\omega}{dt} \tag{3.7}$$

The direction of α is the same as $d\theta$ and ω. If ω decreases, α is called an angular deceleration and therefore has a direction opposite to ω. From Equations 3.6 and 3.7, by eliminating dt we obtain

$$\alpha \, d\theta = \omega \, d\omega \tag{3.8}$$

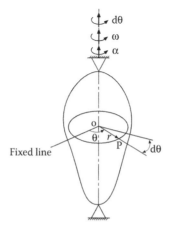

FIGURE 3.3 Fixed-axis rotation.

3.2.2.2 Tangential and Normal Components of Fixed-Axis Rotation

Consider point "O" on the rotational axis shown in Figure 3.4. Point P rotates with radius r along a circular path. The magnitude of the tangential velocity V_t which is tangent at point P is given as

$$V_t = \omega r \tag{3.9}$$

or, in vector notation, as

$$\bar{V}_t = \bar{\omega} \times \bar{r} \tag{3.10}$$

Equation 3.10 provides the direction of velocity $\bar{V}_t$ as well as the magnitude. As shown in Figure 3.5, the acceleration of point P has two components: normal and tangential. The tangential acceleration component a_t is given as

$$a_t = \frac{dV_t}{dt} \tag{3.11}$$

where $V_t = \omega r$. In Equation 3.12, since the position vector, r, of point P is constant

$$\frac{dr}{dt} = 0$$

$$a_t = \frac{d}{dt}(\omega r) = \frac{d\omega}{dt}r + \omega\frac{dr}{dt} \tag{3.12}$$

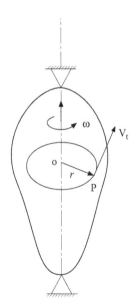

FIGURE 3.4 Tangential velocity.

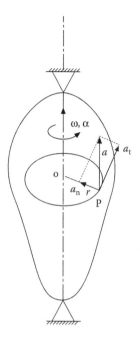

FIGURE 3.5 Acceleration components.

$$a_t = \frac{d\omega}{dt} r = \alpha r \qquad (3.13)$$

Equation 3.13 gives the change in magnitude of the velocity. The normal acceleration component a_n is given as

$$a_n = \frac{V_t^2}{r} \qquad (3.14)$$

Substituting $V_t = \omega r$ into Equation 3.14 yields

$$a_n = \omega^2 r \qquad (3.15)$$

Using Equation 3.10, acceleration in vector form can be written as

$$\bar{a} = \frac{d\bar{V_t}}{dt} = \frac{d}{dt}(\bar{\omega} \times \bar{r})$$

$$\bar{a} = \frac{d\bar{\omega}}{dt} \times \bar{r} + \bar{\omega} \times \frac{d\bar{r}}{dt}$$

$$= \bar{\alpha} \times \bar{r} + \bar{\omega} \times \bar{V_t}$$

or

$$= \bar{\alpha} \times \bar{r} + \bar{\omega} \times (\bar{\omega} \times \bar{r})$$

Note that $d\bar{r}/dt \uparrow 0$ because its direction changes with time. Defining tangential and normal acceleration as

$$\bar{a}_t = \bar{\alpha} \times \bar{r}$$

and

$$\bar{a}_n = \bar{\omega} \times \left(\bar{\omega} \times \bar{r} \right)$$

the expression of acceleration in vector form becomes

$$\bar{a} = \bar{a}_t + \bar{a}_n \tag{3.16}$$

Assuming that a and α are constant, the following relationships can be used to transform equations from translational rectilinear motion to fixed-axis rotational motion:

$$s = r\theta$$
$$\alpha = r\omega$$
$$a_t = r\alpha$$
$$a_n = r\omega$$

Complete transformed equations are given in Table 3.1 where S_0 and V_0 are the initial distance and the initial velocity at time zero, respectively. θ_0 and ω_0 are the initial position and angular initial velocity at time zero, respectively.

Example 3.1

For the pulley system shown in Figure 3.6, load A has a constant acceleration of 8 ft/s2 and an initial velocity of 18 ft/s. If the directions of velocity and acceleration of load A are upward, determine:

a. The number of revolutions of the pulley in 5 s.
b. The velocity and position of load B when $t = 5$ s.
c. The acceleration of point C on the pulley when $t = 0$.

TABLE 3.1
Transformation from Rectilinear Motion to Rotational Motion

Rectilinear Motion	Rotational Motion
$V = V_0 + at$	$\omega = \omega_0 + \alpha t$
$S = S_0 + V_0 t + \dfrac{1}{2} a t^2$	$\theta = \theta_0 + \omega_0 t + \dfrac{1}{2} \alpha t^2$
$V^2 = V_0^2 + 2a(S - S_0)$	$\omega^2 = \omega_0^2 + 2\alpha(\theta - \theta_0)$

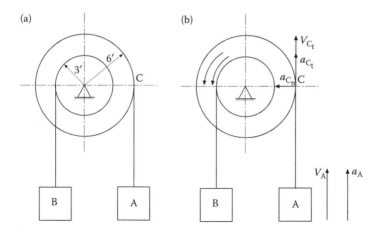

FIGURE 3.6 (a) Pulley system and (b) motion of the pulley system.

SOLUTION

a. The motion of the pulley system is shown in Figure 3.6b. Assuming the wires carrying the loads are inextensible

$$(V_C)_{t=0} = (V_A)_{t=0} = 18 \text{ ft/s}$$

at time zero. Also, acceleration at $t = 0$ is

$$(a_C)_t = a_A = 8 \text{ ft/s}^2$$

Position of point C or A at time t is

$$\theta = \theta_0 + \omega_0 t + \frac{1}{2}\alpha t^2$$

From

$$V_0 = r\omega_0 \implies \omega_0 = \frac{V_0}{r} = \frac{18}{6} = 3 \text{ rad/s}$$

and

$$a_0 = r\alpha \implies \alpha = \frac{a_0}{r} = \frac{8}{6} = \frac{4}{3} \text{ rad/s}$$

Assuming initial position $\theta_0 = 0$, position of point C and A is

$$\theta = 3 \times (5) + \frac{1}{2}\left(\frac{4}{3}\right)(5)^2 = 31.66 \text{ rad}$$

and the number of revolutions is

$$31.66 \text{ rad} \times \frac{\text{rev}}{2\pi} = 5.04 \text{ rev}$$

b. The velocity of point B is

$$V_B = r\omega$$

where

$$\omega = \omega_0 + \alpha t$$

and substituting known values yields

$$\omega = 3 + \left(\frac{4}{3}\right)(5) = 9.66 \text{ rad/s}$$

Then

$$V_B = 3 \times 9.66 = 28.98 \text{ ft/s}$$

$$S_B = r\theta = 3 \times (31.66) = 94.98 \text{ ft}$$

c. The acceleration of point C on the pulley when $t = 0$ is

$$(a_C)_n = r\omega_0^2$$

$$(a_C)_n = 6 \times (3)^2 = 54 \text{ ft/s}^2$$

Knowing that

$$(a_C)_t = 8 \text{ ft/s}^2$$

$$a_C = \sqrt{a_t^2 + a_n^2}$$

$$a_C = \sqrt{8^2 + 54^2} = 54.59 \text{ ft/s}^2$$

and

$$\tan\varphi = \frac{a_t}{a_n} = \frac{8}{54} = 8.4°$$

3.2.3 GENERAL PLANE MOTION

If the resultant force system on a body consists of both a couple and a force passing through the center of mass as shown in Figure 3.7, the body will show both

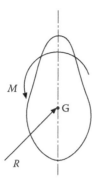

FIGURE 3.7 General motion.

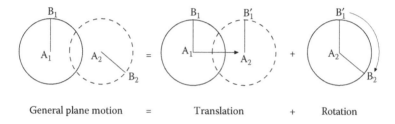

General plane motion = Translation + Rotation

FIGURE 3.8 General motion: translation and rotation.

translational and fixed-axis rotational motion as shown in Figure 3.8. If the applied forces are in the same plane, the motion is referred to as general plane motion or coplanar motion; when the applied forces are not in the same plane, the motion is said to be three dimensional.

3.2.4 ABSOLUTE AND RELATIVE VELOCITY IN PLANE MOTION

The concept discussed in this section can be used to determine the absolute velocity of any point relative to any other point on a rigid body. For example, the absolute velocity V_B of point B on a rigid body can be written as

$$\bar{V}_B = \bar{V}_A + \bar{V}_{B/A} \tag{3.17}$$

where V_A is the absolute velocity of point A (translation) and $V_{B/A}$ is the relative velocity of point B with respect to point A (i.e., rotation of point B about point A). In other words, the absolute velocity of point B is a combination of the translation of point A and the rotation of point B about point A. In vector notation, the rotation of point B about point A is given by

$$\bar{V}_{B/A} = \bar{\omega} \times \bar{r}_{B/A} \tag{3.18}$$

where $\bar{\omega}$ is the absolute angular velocity of the rigid body and $\bar{r}_{B/A}$ is the relative position vector.

Example 3.2

The three-bar link OA shown in Figure 3.9 has a counterclockwise angular velocity of 12 rad/s during a short interval of motion. When link CB is vertical, point A has coordinates (−50 mm, 75 mm). Determine the angular velocity of links AB and BC.

SOLUTION

Consider link AB. The absolute velocity of point A is

$$\bar{V}_A = \bar{V}_B + \bar{V}_{A/B} \tag{a}$$

From link OA, we can determine

$$\bar{V}_A = \bar{V}_O + \bar{V}_{A/O}$$

Since point O is fixed, the velocity of point O is zero. Then

$$\bar{V}_A = \bar{V}_{A/O} = \bar{\omega}_{OA} \times \bar{r}_{A/O}$$

The relative position vector, $\bar{r}_{A/O} = \bar{r}_{OA}$, is

$$\bar{r}_{OA} = -0.05\bar{i} + 0.075\bar{j}$$

Substituting yields

$$\bar{V}_A = \bar{\omega}_{OA} \times \bar{r}_{A/O} = \left(12\bar{k}\right) \times \left(-0.05\bar{i} + 0.075\bar{j}\right)$$

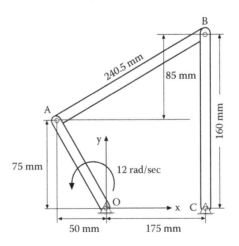

FIGURE 3.9   Three-bar link.

or

$$\bar{V}_A = -0.6\bar{j} - 0.9\bar{i} \tag{b}$$

The absolute velocity of point B with respect to fixed point C is

$$\bar{V}_B = \bar{V}_C + \bar{V}_{B/C}$$

Since $\bar{V}_C = 0$

$$\bar{V}_B = \bar{V}_{B/C} = \bar{\omega}_{BC} \times \bar{r}_{B/C}$$

where

$$\bar{r}_{B/C} = \bar{r}_{CB} = 0.16\bar{j}$$

Then

$$\bar{V}_B = \omega_{BC}\bar{k} \times \left(0.16\bar{j}\right)$$

or

$$\bar{V}_B = -0.16\omega_{BC}\bar{i} \tag{c}$$

The last term of Equation (a) is

$$\bar{V}_{A/B} = \bar{\omega}_{AB} \times \bar{r}_{A/B}$$

The relative position vector $\bar{r}_{A/B}$ is

$$\bar{r}_{A/B} = \bar{r}_{BA} = -0.225\bar{i} - 0.085\bar{j}$$

$$\bar{V}_{A/B} = \omega_{AB}\,\bar{k} \times \left(-0.225\bar{i} - 0.085\,\bar{j}\right)$$

$$\bar{V}_{A/B} = -0.225\omega_{AB}\bar{j} + 0.085\omega_{AB}\bar{i} \tag{d}$$

Substituting Equations (b) through (d) into Equation (a) gives

$$\left(-0.6\,\bar{j} - 0.9\bar{i}\right) = \left(-0.16\omega_{BC}\bar{i}\right) + \left(-0.225\omega_{AB}\bar{j} + 0.085\omega_{AB}\bar{i}\right)$$

From the $\bar{j}$ component

$$-0.6 = -0.225\omega_{AB}$$

$$\omega_{AB} = \frac{-0.6}{-0.225} = 2.67 \text{ rad/s}$$

From the $\bar{i}$ component

$$-0.9 = -0.16\omega_{BC} + 0.085\omega_{AB}$$

or

$$-0.9 = -0.16\omega_{BC} + (0.085)\,(2.67)$$

Then

$$\omega_{BC} = \frac{0.9 + 0.22695}{0.16} = 7\,\text{rad/s}$$

Example 3.3

In the engine system shown in Figure 3.10, determine the velocity of the piston and the angular velocity of connecting rod AB. Assume that when crank angle $\alpha = 30°$, crank arm CB rotates at 2500 rpm.

SOLUTION

The angular velocity of arm BC is

$$\omega_{BC} = \frac{2500 \times 2\pi}{60} = 262\,\text{rad/s}$$

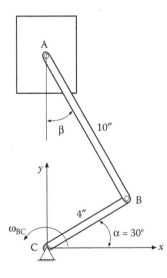

FIGURE 3.10 Engine system.

From the sine law

$$\frac{4}{\sin\beta} = \frac{10}{\sin 60}$$

$$\sin\beta = \frac{4\sin 60}{10}$$

Then

$$\beta = 20.3°$$

The absolute velocity of point A (piston) is

$$\bar{V}_A = \bar{V}_B + \bar{V}_{A/B} \qquad \text{(a)}$$

but, from crank CB,

$$\bar{V}_B = \bar{V}_C + \bar{V}_{B/C} = \bar{V}_{B/C} = \bar{\omega}_{CB} \times \bar{r}_{CB} \qquad \text{(b)}$$

The relative position vector, $\bar{r}_{B/C} = \bar{r}_{CB}$, is

$$\bar{r}_{CB} = 4\cos 30\bar{i} + 4\sin 30\bar{j}$$

$$= 3.46\bar{i} + 2\bar{j}$$

substituting which into Equation (b) yields

$$\bar{V}_B = 262\bar{k} \times \left(3.46\bar{i} + 2\bar{j}\right)$$

$$\bar{V}_B = 906.5\bar{j} - 524\bar{i} \qquad \text{(c)}$$

The velocity of point A relative to point B is

$$\bar{V}_{A/B} = \bar{\omega}_{AB} \times \bar{r}_{A/B} \qquad \text{(d)}$$

where

$$\bar{r}_{A/B} = \bar{r}_{BA} = -10\sin 20.3\bar{i} + 10\cos 20.3\bar{j}$$

substituting which into Equation (d) gives

$$\bar{V}_{A/B} = \omega_{AB}\bar{k} \times \left(-3.47\bar{i} + 9.38\bar{j}\right)$$

$$\bar{V}_{A/B} = -3.47\omega_{AB}\bar{j} - 9.38\omega_{AB}\bar{i} \qquad \text{(e)}$$

Note that the piston only moves in the y-direction. Thus, from Equation (a)

$$V_A \bar{j} = \left(906.5\bar{j} - 524\bar{i}\right) + \left(-3.47\omega_{AB}\bar{j} - 9.38\omega_{AB}\bar{i}\right)$$

From the $\bar{i}$ component

$$-524 - 9.38\omega_{AB} = 0$$

$$\omega_{AB} = \frac{-524}{9.38} = -55.9 \text{ rad/s}$$

From the $\bar{j}$ component

$$V_A = 906.5 - 3.47\omega_{AB}$$

$$= 906.5 - 3.47\,(-55.9) = 1100.5 \text{ in./s}$$

or

$$V_A = 91.7 \text{ ft/s}$$

3.2.5 INSTANTANEOUS CENTER OF ZERO VELOCITY (IC)

Consider the rigid body shown in Figure 3.11 in which the velocity directions of point A and B are given at any instant of time. The intersection point (IC) of two lines passing through points A and B, which is perpendicular to V_A and V_B, respectively, is the center of rotation of the rigid body and has zero velocity. Point IC is on the instantaneous axis of rotation which is perpendicular to the plane of the paper and is called the instantaneous center of zero velocity. In general, this point may be on or off a planar rigid body.

From Figure 3.11, the absolute velocity of point A can be determined by

$$\bar{V}_A = \bar{V}_C + \bar{V}_{A/C} \tag{3.19}$$

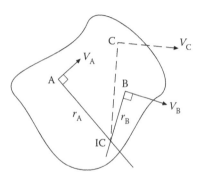

FIGURE 3.11 Instantaneous center of zero velocity.

Since the velocity of point C is zero, Equation 3.19 can be simplified to

$$\bar{V}_A = \bar{V}_{A/C} = \bar{\omega}_{AC} \times \bar{r}_A \qquad (3.20)$$

Note that zero velocity does not mean that point IC has zero acceleration.

3.2.5.1 Determining IC

Case I: As shown in Figure 3.12a, if the velocity directions of two points on a rigid body are known, simply draw a perpendicular line to the velocity directions to find the intersection point IC. Note that if IC is known, velocity of any arbitrary point on the rigid body can be determined.

Case II: As shown in Figure 3.12b, if the velocity directions of two points are parallel and the unequal velocity magnitudes are known, the IC lies on the common line perpendicular to these velocity directions and is located by direct proportion.

Example 3.4

Locate the IC of member BC of the three-link mechanism shown in Figure 3.13. Determine the angular velocity of BC if the angular velocity of BA is 2 rad/s clockwise (CW).

SOLUTION

$$\bar{V}_B = \bar{V}_A + \bar{V}_{B/A} = 0 + \bar{\omega}_{BA} \times \bar{r}_{B/A} = \left(-2\bar{k}\right) \times \left(1.5\bar{j}\right) = 3.0\bar{i}$$

But

$$\bar{V}_B = \bar{\omega}_{CB} \times \bar{r}_{IC\text{-}B} = \left(\omega_{CB}\bar{k}\right) \times \left(-15\bar{j}\right)$$

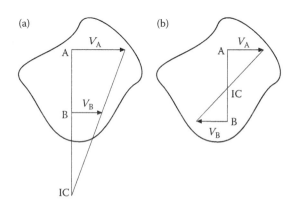

FIGURE 3.12   Determining the instantaneous center of zero velocity. (a) Velocity directions of two points are known, (b) velocity directions of two points are parallel and the unequal velocity magnitudes are known.

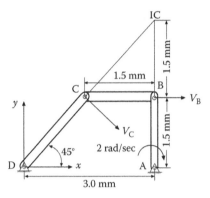

FIGURE 3.13 Three-link mechanism.

Then

$$3.0\bar{i} = \left(\omega_{CB}\bar{k}\right) \times \left(-1.5\bar{j}\right) = 1.5\omega_{CB}\bar{i}$$

$$\omega_{CB} = 2 \text{ rad/s}$$

3.2.6 ABSOLUTE AND RELATIVE ACCELERATION IN PLANE MOTION

Similar to equations used for velocity, the absolute acceleration of any point relative to any other point on a rigid body can be written as

$$a_B = a_A + a_{B/A} \qquad (3.21)$$

where a_A and a_B are absolute acceleration of points A and B, respectively. $a_{B/A}$ is the relative acceleration of point B relative to point A. Relative acceleration has two components: tangential and normal.

$$a_{B/A} = (a_{B/A})_t + (a_{B/A})_n \qquad (3.22)$$

where the relative tangential component is given by

$$(a_{B/A})_t = \alpha r_{B/A} \qquad (3.23)$$

and the relative normal component is given by

$$(a_{B/A})_n = \omega^2 r_{B/A} \qquad (3.24)$$

where α and ω are the absolute angular acceleration and angular velocity of the rigid body, respectively. In vector form

$$\left(\bar{a}_{B/A}\right)_t = \bar{\alpha}_{AB} \times \bar{r}_{B/A} \qquad (3.25)$$

and

$$\left(\bar{a}_{B/A}\right)_n = \bar{\omega}_{AB} \times \left(\bar{\omega}_{AB} \times \bar{r}_{B/A}\right) \tag{3.26}$$

Example 3.5

Bar AB of the slider-crank mechanism shown in Figure 3.14 has an angular velocity of 3 rad/s clockwise and a clockwise angular acceleration of 6 rad/s². Determine the angular acceleration of bar BC.

SOLUTION

The absolute velocity of point B is

$$\bar{V}_B = \bar{V}_A + \bar{V}_{B/A}$$

where $\bar{V}_A = 0$. Then

$$\bar{V}_B = \bar{V}_{B/A} = \bar{\omega}_{AB} \times \bar{r}_{AB} \tag{a}$$

where

$$\bar{r}_{AB} = 0.4\,\bar{j} \quad \text{and} \quad \bar{\omega}_{AB} = -3\bar{k}$$

Substituting these known values into Equation (a) yields

$$\bar{V}_B = \left(-3\bar{k}\right) \times \left(0.4\,\bar{j}\right) = 1.2\bar{i}$$

The absolute velocity of point C is

$$\bar{V}_C = \bar{V}_B + \bar{V}_{C/B} \tag{b}$$

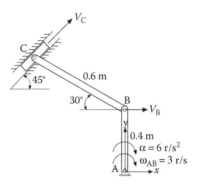

FIGURE 3.14 Slider-crank mechanism.

Therefore,

$$\bar{V}_C = \bar{V}_B + \left(\omega_{BC}\bar{k}\right) \times \left(\bar{r}_{BC}\right)$$

$$\bar{V}_C = 1.2\bar{i} + \left(\omega_{BC}\bar{k}\right) \times \left(-0.6\cos 30\bar{i} + 0.6\sin 30\bar{j}\right)$$

$$\bar{V}_C = 1.2\bar{i} - 0.3\omega_{BC}\bar{i} - 0.52\omega_{BC}\bar{j}$$

where components of $\bar{V}_C$ are

$$\bar{V}_C = V_C\left(\cos 45\bar{i} + \cos 45\bar{j}\right)$$

$$\bar{V}_C = V_C\left(0.707\bar{i} + 0.707\bar{j}\right)$$

Substituting these known values into Equation (b) yields

$$0.707\ V_C\bar{i} + 0.707\ V_C\bar{j} = 1.2\bar{i} - 0.3\omega_{BC}\bar{i} - 0.52\omega_{BC}\bar{j}$$

From the $\bar{i}$ and $\bar{j}$ components, we have

$$0.707V_C = 1.2 - 0.3\omega_{BC} \tag{c}$$

$$0.707V_C = -0.52\omega_{BC} \tag{d}$$

Solving Equations (c) and (d) simultaneously yields

$$\omega_{BC} = -5.45 \text{ rad/s}$$

The absolute acceleration of point B is

$$\bar{a}_B = \bar{a}_A + \bar{a}_{B/A}$$

where, since $\bar{a}_A = 0$,

$$\bar{a}_B = \bar{a}_{B/A} = \bar{\alpha}_{AB} \times \bar{r}_{B/A} + \bar{\omega}_{AB} \times \left(\bar{\omega}_{AB} \times \bar{r}_{B/A}\right)$$

where relative position $\bar{r}_{B/A} = \bar{r}_{AB} = 0.4\,\bar{j}$, $\bar{\alpha} = -6\bar{k}$ and $\bar{\omega}_{AB} = -3\bar{k}$
Thus, substituting known values yields

$$\bar{a}_B = \left(-6\bar{k}\right) \times \left(0.4\bar{j}\right) + \left(-3\bar{k}\right) \times \left[\left(-3\bar{k}\right) \times \left(0.4\bar{j}\right)\right]$$

$$\bar{a}_B = 2.4\bar{i} - 3.6\,\bar{j}$$

Now consider the acceleration of point C:

$$\bar{a}_C = \bar{a}_B + \bar{a}_{C/B}$$

$$\bar{a}_C = \bar{a}_B + \bar{\alpha}_{CB} \times \bar{r}_{C/B} + \bar{\omega}_{CB} \times \left(\bar{\omega}_{CB} \times \bar{r}_{C/B}\right)$$

Note that $\bar{r}_{C/B} = \bar{r}_{BC} = -0.6\cos 30\,\bar{i} + 0.6\sin 30\,\bar{j}$ where components of a_c are

$$\bar{a}_c = a_c\left(0.707\bar{i} + 0.707\bar{j}\right)$$

Substituting known values yields

$$0.707a_C\bar{i} + 0.707a_C\bar{j} = \left(2.4\bar{i} - 3.6\,\bar{j}\right) + \left[\alpha_{CB}\bar{k} \times \left(-0.6\cos 30\,\bar{i} + 0.6\sin 30\,\bar{j}\right)\right]$$
$$+ \left(-5.45\bar{k}\right) \times \left[\left(-5.45\,\bar{k}\right) \times \left(-0.6\cos 30\,\bar{i} + 0.6\sin 30\,\bar{j}\right)\right]$$

or

$$0.707a_C\bar{i} + 0.707a_C\bar{j} = \left(2.4\bar{i} - 3.6\,\bar{j}\right) + \left(-0.52\alpha_{CB}\bar{j} - 0.3\alpha_{CB}\bar{i}\right) + \left(15.42\bar{i} - 8.94\,\bar{j}\right)$$

Rearranging yields

$$0.707a_C\bar{i} + 0.707a_C\bar{j} = \left(17.82 - 0.3\alpha_{CB}\right)\bar{i} + \left(-12.54 - 0.52\alpha_{CB}\right)\bar{j}$$

From the $\bar{i}$ and $\bar{j}$ components, the scalar equations can be written as

$$0.707a_C = 17.82 - 0.3\alpha_{CB} \qquad \text{(e)}$$

and

$$0.707a_C = -12.54 - 0.52\alpha_{CB} \qquad \text{(f)}$$

From Equations (e) and (f)

$$17.82 - 0.3\alpha_{CB} = -12.54 - 0.52\alpha_{CB}$$

and

$$\alpha_{CB} = -138 \text{ rad/s}^2$$

3.3   KINETICS OF A RIGID BODY

3.3.1   TRANSLATION

When a rigid body undergoes a translation, all the points on the rigid body have the same acceleration as the center of mass. Thus

$$\bar{a}_G = \bar{a} \tag{3.27}$$

The motion of a rigid body in translation is governed by

$$\sum F = m\bar{a}_G \tag{3.28}$$

Equation 3.28 demonstrates Newton's laws of motion. Since the rigid body is not rotating, the sum of the moments of the applied forces about an axis passing through the center of mass must be equal to zero. Hence

$$\sum M_G = 0 \tag{3.29}$$

Equation 3.29 implies that the angular acceleration is equal to zero. Two types of translational motion-rectilinear and curvilinear-have been defined earlier.

In rectilinear translation, motion is governed by

$$\sum F_x = m\left(a_G\right)_x$$
$$\sum F_y = m\left(a_G\right)_y \tag{3.30}$$
$$\sum M_G = 0$$

As shown in Figure 3.15, it is convenient to use normal and tangential components of acceleration in the equation of motion for curvilinear translation

$$\sum F_n = m\left(a_G\right)_n$$
$$\sum F_t = m\left(a_G\right)_t \tag{3.31}$$
$$\sum M_G = 0$$

The following moment equation can also be used to simplify the analysis:

$$\sum M_A = \sum\left(M_k\right)_A \tag{3.32}$$

Point A in Equation 3.32 is usually located at the intersection of the lines of action of applied forces, where • M_A is the moment about point A on the free-body diagram (FBD) and • $(M_k)_A$ is the moment about point A on the kinetic diagram (KD).

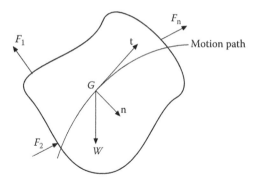

FIGURE 3.15   Curvilinear translation.

Example 3.6

Find the tension in the massless cord which connects the two weights as shown in Figure 3.16. Assume applied force $P = 10$ N and $\mu_k = 0.25$ for both blocks.

SOLUTION

The first step to solve the problem is to draw both the FBD and the KD. The latter shows the equivalent force–couple system with the resultant force applied through the center of mass, G.

a. Draw an FBD and a KD for the first weight as shown Figure 3.17a. Note that since the motion is rectilinear, the KD includes only $F_1 = m_1 a_G$. Applying governing equations in the y-direction yields

$$\sum F_y = m_1 a_y$$

Hence,

$$-96.6 + N_1 = 0$$

and

$$N_1 = 96.6 \text{ N} \tag{a}$$

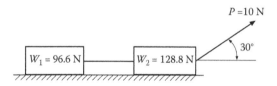

FIGURE 3.16   Two-weight system.

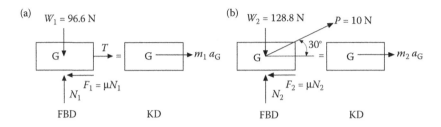

FIGURE 3.17 (a) FBD and KD for the first weight and (b) FBD and KD for the second weight.

Applying governing equations in the x-direction yields

$$\sum F_x = m_1 a_x$$

$$-0.25N_1 + T = m_1 a_x$$

$$T = \frac{96.6}{9.81} a_x + (0.25)(96.6)$$

Then

$$T = 9.85 a_x + 24.15$$

or

$$a_x = \frac{T - 24.15}{9.85} \tag{b}$$

b. Draw an FBD and a KD for the second weight as shown in Figure 3.17b.

$$\sum F_y = m a_y \tag{c}$$

$$-128.8 + N_2 + 10\sin 30 = 0 \Rightarrow N_2 = 123.8\,\text{N}$$

$$\sum F_x = m a_x$$

$$-T - 0.25N_2 + 10\cos 30 = m_2 a_x \tag{d}$$

From Equations (c) and (d), we have

$$T = -0.25(123.8) + 10\cos 30 - m_2\, a_x$$

Substituting Equation (b) yields

$$T = -30.95 + 8.66 - \frac{128.8}{9.81}\left(\frac{T - 24.15}{9.85}\right)$$

$$T = -30.95 + 8.66 - 1.33T + 32.2$$

and

$$T = 4.25\,\text{N}$$

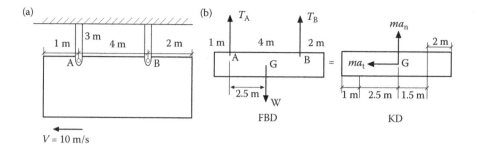

FIGURE 3.18   (a) Swinging bar. (b) Free-body diagram (FBD) and kinetic diagram (KD).

Example 3.7

As shown in Figure 3.18a, the 4 kg swinging bar has a velocity of 10 m/s at some instant of time. Determine the tension in the supporting cables.

SOLUTION

Draw an FBD and a KD as shown in Figure 3.18b. Note that since the motion is curvilinear, the KD includes normal and tangential force components. As seen from the KD, the direction of the tangential force component is in the direction of the motion and the direction of the normal force component is always toward the center. Applying governing equations in the normal direction yields

$$\sum M_A = \sum \left(M_{KD} \right)_A$$

$$4T_B - 2.5W = 2.5Ma_n \quad \Rightarrow \quad T_B = \frac{2.5}{4}\left\{ W + M\frac{V^2}{r} \right\}$$

$$T_B = \frac{2.5}{4}\left\{ 4 \times 9.81 + 4 \times \frac{10^2}{3} \right\}$$

$$T_B = 107.9 \text{ N}$$

and

$$\sum M_B = \sum \left(M_{KD} \right)_B$$

$$-4T_A + 1.5W = -1.5Ma_n \quad \Rightarrow \quad T_A = \frac{1.5}{4}\left\{ W + M\frac{V^2}{r} \right\}$$

$$T_A = \frac{1.5}{4}\left\{ 4 \times 9.81 + 4 \times \frac{10^2}{3} \right\}$$

$$T_A = 64.7 \text{ N}$$

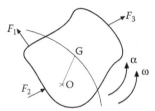

FIGURE 3.19   Rigid-body rotation.

3.3.2   FIXED-AXIS ROTATION

As shown in Figure 3.19, consider a rigid body rotating about the axis passing through point O, which is perpendicular to the plane of motion. As seen from Figure 3.19, while the body rotates about the axis passing through point O, the center of mass, G, moves on a circular path. The FBD and KD of the rotating body are shown in Figure 3.20.

The governing equations are given by

$$\sum F_n = m(a_G)_n = m\omega^2 r \tag{3.33}$$

where

$$(a_G)_n = \frac{V^2}{r} = \frac{(\omega r)^2}{r} = \omega^2 r$$

And

$$\sum F_t = m(a_G)_t = m\alpha r \tag{3.34}$$

where

$$(a_G)_t = \frac{d}{dt}(V) = \frac{d}{dt}(\omega r) = \alpha r$$

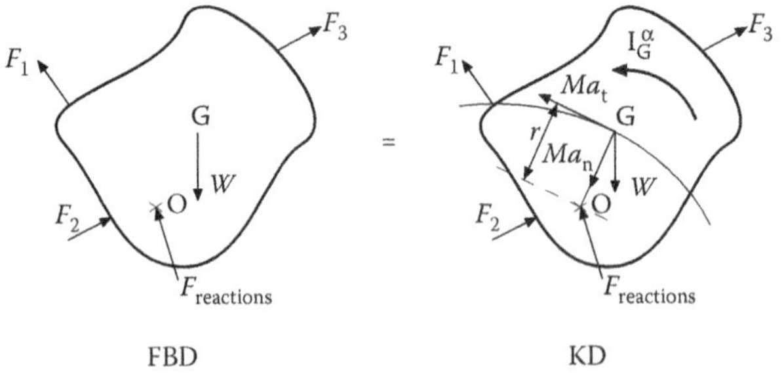

FBD   KD

FIGURE 3.20   Free-body diagram (FBD) and kinetic diagram (KD) of rotating body.

And

$$\sum M_G = I_G \alpha \qquad (3.35)$$

where I_G is the mass moment of inertia.

Equation 3.35 defines the rotational motion. In some cases, the moment equation may be replaced by a moment summation about any arbitrary point, say, A:

$$M_A = I_A \alpha \qquad (3.36)$$

where I_A is the mass moment of inertia about point A. For different geometrical shapes, a mass moment of inertia is given in Table A1.12 (see Appendix 1). If the radius of gyration, k, is given, then the mass moment of inertia can be obtained by

$$I = mk^2 \qquad (3.37)$$

In many cases, an arbitrary point is taken at the pin joints in order to eliminate the unknown forces. Note that if a complete solution cannot be obtained from the equation of motion, the following kinematics equations should be used:

a. If α is time dependent, then

$$\alpha = \frac{d\omega}{dt} \quad \text{and} \quad \alpha\, d\theta = \omega\, d\omega \qquad (3.38)$$

where

$$\omega = \frac{d\theta}{dt}$$

b. If α is constant

$$\omega = \omega_0 + \alpha_c t$$
$$\theta = \theta_0 + \omega_0 t + \tfrac{1}{2}\alpha_c t^2 \qquad (3.39)$$
$$\omega^2 = \omega_0{}^2 + 2\alpha_c \left(\theta - \theta_0\right)$$

c. Use the correlation between linear and angular displacement as shown Figure 3.21. The figure shows a pulley free to rotate about point O under the action of weight W.
   Displacement S is

$$S = r\theta \qquad (3.40)$$

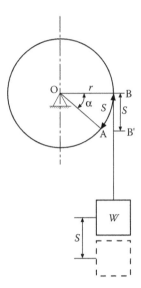

FIGURE 3.21 Correlation between linear and angular displacement.

The derivative of Equation 3.40 yields (note that r is constant)

$$V_t = \frac{ds}{dt} = r\frac{d\theta}{dt} = r\omega$$

$$V_t = r\omega$$

(3.41)

And the acceleration is

$$a_t = r\frac{d\omega}{dt} = r\alpha$$

(3.42)

Example 3.8

The weight of a pulley system shown in Figure 3.22 is 200 lb and the moment of inertia about the center of mass is $I_G = 40$ ft^4. If the weights of the blocks attached to the pulley are 150 and 300 lb, determine the tension in each cord wrapped around the pulleys.

SOLUTION

Figure 3.23 shows the FBD and KD of the pulley system. Since the center of the pulley, point G, is not moving, apply only the following moment equation (see Figure 3.23a):

$$M_G = I_G\alpha \text{ (assuming counter clockwise moments are positive)}$$

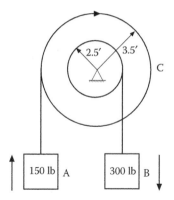

FIGURE 3.22 Pulley system.

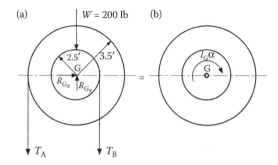

FIGURE 3.23 (a) Free-body diagram (FBD) and (b) kinetic diagram (KD).

Using the FBD and taking the moment about the centroid of the pulley yields

$$-3.5T_A + 2.5T_B = I_G\alpha$$

$$-3.5T_A + 2.5T_B = 40\alpha \tag{a}$$

The FBD of block A is shown in Figure 3.24a. Note that the motion of the mass is in the positive direction.

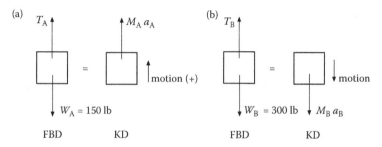

FIGURE 3.24 (a) Free-body diagram (FBD) and (b) kinetic diagram (KD) of A and B blocks.

From the FBD shown in Figure 3.24a

$$\sum F_t = M_A a_A$$

where

$$a_A = \alpha r_A$$

Then

$$\sum F_t = M_A \alpha r_A$$

where $r_A = 3.5$ and $M_A = W_A/g$.
Substituting known values yields

$$T_A - 150 = \frac{150}{g}(3.5\alpha) \text{ assuming positive } y \text{ direction is } \uparrow +$$

$$T_A = \frac{150}{32.2}3.5\alpha + 150$$

or

$$T_A = 16.3\alpha + 150 \qquad\qquad (b)$$

Using the FBD of block B shown in Figure 3.24b

$$\sum F_t = M_B \alpha r_B$$

$$-T_B + 300 = \frac{300}{g}(\alpha \times 2.5) \text{ Assuming positive } y \text{ direction is } \downarrow +$$

Rearranging

$$-T_B = 2.5\alpha\frac{300}{32.2} - 300$$

$$T_B = -2.5\alpha\frac{300}{32.2} + 300$$

$$T_B = -23.3\alpha + 300 \qquad\qquad (c)$$

Substituting Equations (b) and (c) into Equation (a)

$$-3.5(16.3\alpha + 150) + 2.5(-23.3\alpha + 300) = 40\alpha$$

Solving for α

$$-57.05\alpha - 525 - 58.25\alpha + 750 = 40\alpha$$

$$155.3\alpha = 225 \quad \Rightarrow \quad \alpha = 1.45 \text{rad/s}^2$$

From Equations (b) and (c), the tensions about point A, T_A, and point B, T_B, are

$$T_A = 16.3\alpha + 150 = 16.3 \times 1.45 + 150 = 173.6 \text{lb}$$

$$T_B = -23.3\alpha + 300 = -23.3 \times 1.45 + 300 = 266.2 \text{lb}$$

Example 3.9

As shown in Figure 3.25a, a rod with a diameter of 0.02 m is pinned at its end and has an angular velocity of $\omega = 4$ rad/s when it is horizontal. The density of the rod material is 7.8×10^3 kg/m³. Determine the rod's angular acceleration and the pin reaction at this instant.

SOLUTION

Draw an FBD and KD of the rolling cylinder as shown in Figure 3.25b.
The mass of the rod is

$$m = 7.8 \times 10^3 \times \frac{\pi(0.02)^2}{4} \times 1.4 = 3.43 \text{ kg}$$

From the governing equations

$$\text{assuming positive } x \text{ direction is } \rightarrow + \quad \sum F_x = m(a_G)_x$$

$$A_x = -m(\omega^2 r) = -(3.43)(4^2)(0.7) = -38.4 \text{ N}$$

(a)

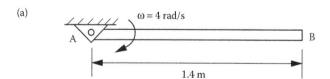

(b)

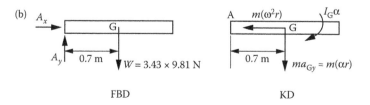

FBD   KD

FIGURE 3.25 (a) Pinned end rod. (b) Free-body diagram (FBD) and kinetic diagram (KD).

Then

$$A_x = 38.4 \text{ N} \leftarrow$$

Angular acceleration of the rod is calculated by

$$\Sigma M_A = I_A \alpha \quad \text{(assuming counterclockwise moments are positive)}$$

Mass moment of inertia of the rod is

$$I_A = \frac{1}{3}ml^2 = \frac{1}{3}(3.43)(1.4)^2 = 2.24 \text{ Nms}^2$$

Substituting known values into the moment equation yields

$$(3.43 \times 9.81)0.7 = 2.24\alpha$$

and

$$\alpha = 10.5 \text{ rad/s}^2$$

Summation of the forces in the y-direction yields

$$\text{Assuming positive } y \text{ direction is } \uparrow + \sum F_y = m(a_G)_y = m(\alpha r)$$

$$A_y - (3.43 \times 9.81) = -(3.43)(10.5)(0.7)$$

and

$$A_y = 8.44 \text{ N} \uparrow$$

3.3.3 GENERAL PLANE MOTION

In general plane motion, the centroid of a body, G, demonstrates both translational and rotational motion (like rolling bodies). The governing equations are given by

$$\sum F_x = m(a_G)_x \quad \rightarrow \quad \text{translation of G in } x\text{-direction} \qquad (3.43)$$

$$\sum F_y = m(a_G)_y \quad \rightarrow \quad \text{translation of G in the } y\text{-direction} \qquad (3.44)$$

$$\sum M_G = I_G \alpha \quad \rightarrow \quad \text{rotation of center of mass G} \qquad (3.45)$$

or taking the moment about an arbitrary point A

$$\sum M_A = I_A \alpha \qquad (3.46)$$

Note that if the complete solution cannot be obtained from the governing equations, the following kinematics equations should be used:

$$\bar{V}_B = \bar{V}_A + \bar{V}_{B/A}$$
$$\bar{a}_B = \bar{a}_A + \bar{a}_{B/A}$$

(3.47)

3.4 ROLLING PROBLEMS

Assume that a solid cylinder of r radius and W weight rolls under the action of P along the horizontal plane, as shown in Figure 3.26a. The FBD and KD are shown in Figure 3.26b. Rolling problems can be classified into two types.

1. *Rolling without Slipping*: In this case, enough friction force develops between a solid cylinder and the floor, for example, so that the solid cylinder rolls without slipping. Thus, the following equation must be satisfied:

$$F' \geq F$$

(3.48)

where F' is the friction force given as

$$F' = \mu N$$

(3.49)

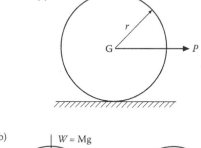

(a)

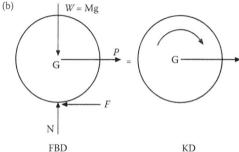

(b)

FBD   KD

FIGURE 3.26 (a) Rolling solid cylinder and (b) free-body and kinetic diagrams.

where μ is the coefficient of friction. In this case, acceleration of the center of mass G can be determined by

$$a_G = r\alpha \qquad (3.50)$$

2. *Rolling with Slipping*: In this case, since enough friction force does not develop ($F' < F$), Equation 3.50 does not apply; that is, a_G and α are independent of each other. In this case, we assume that

$$F = F' \qquad (3.51)$$

and determine the values of a_G and α from the equation of motion.

Example 3.10

The solid cylinder shown in Figure 3.27a weighs 100 lb and rolls in the x-direction. Determine the angular acceleration and the acceleration of the center of mass of the wheel. Assume the radius of gyration of the wheel is $k = 1.5$ ft and the coefficient of friction, μ, between the solid cylinder and the horizontal plane is 0.25.

SOLUTION

Draw an FBD and KD of the rolling cylinder as shown in Figure 3.27b.
The governing equations are

$$\sum F_x = m(a_G)_x$$

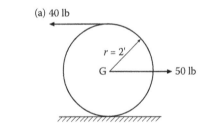

(a) 40 lb

$r = 2'$

G   50 lb

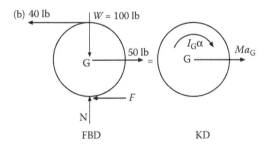

(b) 40 lb $W = 100$ lb

50 lb

G

F

N

FBD

$I_G\alpha$

G Ma_G

KD

FIGURE 3.27 (a) Solid cylinder rolling in the x-direction and (b) free-body and kinetic diagrams.

$$\sum F_y = m(a_G)_y$$

$$\sum M_G = I_G \alpha$$

If we assume no slipping conditions

$$a_G = r\alpha = 2\alpha$$

$$\sum F_x = m(a_G)_x$$

$$50 - 40 - F = m(a_G)_x = \frac{100}{32.2}(2\alpha)$$

$$10 - F = 6.21\alpha$$

$$\alpha = \frac{10 - F}{6.21} \qquad\qquad (a)$$

Applying governing equations in the y-direction

$$\sum F_y = m(a_G)_y$$

Since there is no motion in the y-direction, $a_G = 0$. Thus,

$$\sum F_y = N - 100 = 0 \quad \Rightarrow \quad N = 100\text{lb}$$

$$\sum M_G = I_G \alpha \quad \text{(assuming counterclockwise moments are positive)}$$

$$-40 \times 2 + F \times 2 = I_G \alpha \qquad\qquad (b)$$

where

$$I_G = mk^2 = \frac{100}{32.2} \times 1.5^2 = 6.99$$

and substituting known values into Equation (b) yields

$$2F - 80 = 6.99\alpha \qquad\qquad (c)$$

Substituting Equation (a) into Equation (c) gives

$$2F = 6.99\left(\frac{10 - F}{6.21}\right) + 80$$

and solving for F gives $F = 29.19$ lb

The maximum friction force to prevent slipping is

$$F' = \mu N = 0.25 \times 100 = 25 \text{ lb}$$

Since

$$F' < F$$

there is not enough friction force to prevent slipping. Thus, we assume $F' = F = 25$ lb and apply the governing equation

$$\sum F_x = m(a_G)_x$$

solving for acceleration a_G gives

$$50 - 40 - 25 = \frac{100}{32.2}(a_G)$$

$$a_G = -4.83 \text{ f/s}^2$$

$$\sum M_G = I_G \alpha$$

$$2 \times 25 - 40 \times 2 = 6.99\alpha$$

$$\alpha = \frac{-30}{6.99} = -4.29 \text{ rad/s}^2$$

Note that the negative sign indicates that the motion is deceleration.

3.5   PLANAR KINETIC ENERGY AND WORK

When a rigid body of mass m is subjected to translation and rotation about a fixed axis, the kinetic energy is given by

$$T = \frac{1}{2}mV_G^2 + \frac{1}{2}I_G\omega^2 \tag{3.52}$$

The first term of Equation 3.52 is the kinetic energy due to translation and the second term represents the kinetic energy due to fixed-axis rotation.

A force that is displaced along its line of action will do work. Work can be defined using the following two equations:

a. The work of a spring is

$$W = \frac{1}{2}kx^2 \tag{3.53}$$

where k is the spring constant and x is the spring displacement.

b. The work of a rotating body is

$$W = M \times \Delta\theta \qquad (3.54)$$

where M is the moment of a couple ($M = F \times r$) and $\Delta\theta$ is the angular displacement.

3.5.1 PRINCIPLE OF WORK AND ENERGY

Equation 3.55 states that the total work done on a rigid body as it moves from one position to another is equal to the change in its kinetic energy. This is called the *principle of work and energy*.

$$\sum W_{1-2} = T_2 - T_1 \qquad (3.55)$$

where the initial kinetic energy, T_1, is

$$T_1 = \frac{1}{2}mV_1^2 \qquad (3.56)$$

and the final kinetic energy, T_2, is

$$T_2 = \frac{1}{2}mV_2^2 \qquad (3.57)$$

ΣW_{1-2} is the work done on the rigid body by all the external forces and couples.

Example 3.11

As shown in Figure 3.28a, a solid disk of 10 lb is raised 3 ft from the equilibrium condition by applying force $F = 30$ lb. If the spring is initially unstretched, determine the spring constant k.

SOLUTION

$$\text{Work of weight} \quad \Rightarrow \quad W \times y = -10 \times 3 = -30 \text{ lb-ft}$$

$$\text{Work of force} \quad \Rightarrow \quad F \times y = +30 \times 6 = 180 \text{ lb-ft}$$

$$\text{Work of spring} \quad \Rightarrow \quad -\frac{1}{2}ky^2 = -\frac{1}{2}k(3)^2 = -4.5k$$

Note that if the spring rises 3 ft, force F will rise 6 ft. Using the FBD and applying the principle of work and energy yield

$$\sum W_{1-2} = T_2 - T_1$$

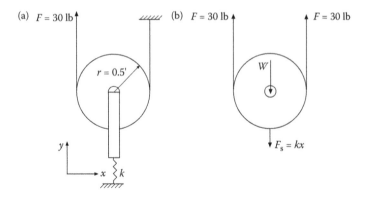

FIGURE 3.28 (a) Disk–spring system and (b) free-body diagram.

Since the solid disk is at rest initially and will also be at rest after the disk is raised, $T_1 = 0$ and $T_2 = 0$, then

$$\sum W_{1-2} = -30 + 180 - 4.5k = 0$$

$$k = \frac{180 - 30}{4.5} = 33.33 \text{ lb/ft}$$

3.5.2 Conservation of Energy

If only conservative forces (such as spring force or gravity force) are applied, then the following conservation of energy theorem can be used; otherwise the work and energy theorem should be used.

$$T_1 + V_1 = T_2 + V_2 \tag{3.58}$$

where T and V are the kinetic and potential energies of the system, respectively.

Example 3.12

A 300-lb weight is attached to a grooved disk by a rope, as shown in Figure 3.29a. Determine the maximum drop of W when the weight is released from rest while the spring is unstretched. Assume a spring constant of 350 lb/ft.

SOLUTION

Since the system has gravity and spring force, the conservation of energy theorem can be used as shown in Equation (a).

$$T_1 + V_1 = T_2 + V_2 \tag{a}$$

Since the weight is at rest initially and finally,

$$T_1 = 0 \quad \text{and} \quad T_2 = 0$$

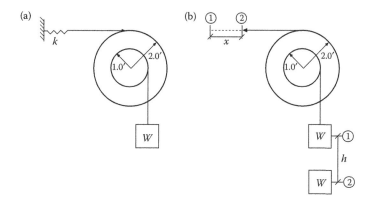

FIGURE 3.29 (a) Grooved-disk system and (b) displacement of the weight W.

From Figure 3.29b,

$$x = r_1\theta = 2 \times \theta \quad \Rightarrow \quad \theta = \frac{x}{2} \tag{b}$$

$$h = r_2\theta = 1 \times \theta \quad \Rightarrow \quad \theta = h \tag{c}$$

From Equations (b) and (c),

$$x = 2h \tag{d}$$

The initial potential energy is due to gravitational force and is determined as

$$V_1 = W \times h = 300 \times h \tag{e}$$

The final potential energy is due to spring force and is determined as

$$V_2 = \frac{1}{2}kx^2 = \frac{1}{2} \times 350 \times (2h)^2 = 700 \times h^2 \tag{f}$$

Substituting Equations (e) and (f) into Equation (a) gives

$$300 \times h = 700 \times h_2 \quad \Rightarrow \quad h = 0.43 \text{ ft}$$

3.5.3 PRINCIPLE OF LINEAR IMPULSE AND MOMENTUM

Equation 3.59 demonstrates the principle of linear impulse and momentum which states that the sum of impulses, $\Sigma \int F \, dt$, created by external forces during a time interval, $\Delta t = t_2 - t_1$, is equal to the change in linear momentum.

$$\sum \int_{t_1}^{t_2} F \, dt = m(V_G)_2 - m(V_G)_1 = L_2 - L_1 \tag{3.59}$$

where the product of mass and velocity is defined as the linear momentum, $L = mV$. For a constant force, Equation 3.59 takes the form

$$F \, \Delta t = m \, \Delta V \tag{3.60}$$

3.5.4 PRINCIPLE OF ANGULAR IMPULSE AND MOMENTUM

Equation 3.61 demonstrates the principle of angular impulse and momentum which states that the sum of the angular impulse, $\Sigma \int M \, dt$, during a time interval is equal to the change in angular momentum, $I \, \Delta \omega$.

$$\sum \int_{t_1}^{t_2} M_G \, dt = I_G \omega_2 - I_G \omega_1 = H_2 - H_1 \tag{3.61}$$

where H is the angular momentum, I_G is the mass moment of inertia about the center of mass, and ω is the angular velocity of the rigid body. If the moment on a rigid body is constant, then Equation 3.61 takes the form

$$M \, \Delta t = I \, \Delta \omega \tag{3.62}$$

Note that the impulse-momentum equations can be used when the calculation of a motion parameters at a specified time is required.

3.5.5 IMPACT

Impact is defined as the collision between two bodies, which generates relatively large contact forces in a very short time interval. If colliding bodies are not subject to external forces, their total linear momentum must be the same before and after impact.

3.5.5.1 Direct Central Impact

As shown in Figure 3.30, if the center of mass of m_1 and m_2 travel along the same straight line with velocity V_1 greater than V_2, collision occurs with the contact forces directed along the line of the center of mass. This is called direct central impact.

Assume that two bodies m_1 and m_2 with velocities V_1 and V_2 collide and reach final velocities of V_1 and V_2, respectively, after the impact. If the effect of external forces is ignored, their total momentum is conserved.

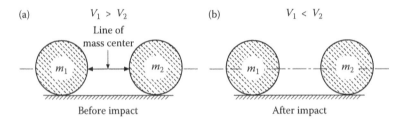

FIGURE 3.30 Impact on a straight line.

$$m_1 V_1 + m_2 V_2 = m_1 V_1' + m_2 V_2' \tag{3.63}$$

To solve the two unknowns, V_1 and V_2, an additional relationship called the *coefficient of restitution*, e, is used:

$$e = \frac{V_2' - V_1'}{V_1 - V_2} = \frac{\text{Relative velocity of separation}}{\text{Relative velocity of approach}} \tag{3.64}$$

If $e = 0$ the impact is said to be perfectly plastic, and if $e = 1$ the impact is said to be perfectly elastic with no energy loss, that is, the total kinetic energy remains the same before and after the impact.

3.5.5.2 Oblique Central Impact

A case of impact where the initial and final velocities of the bodies are not parallel, and the bodies approach each other at an oblique angle, as shown in Figure 3.31, is known as oblique central impact.

To solve the unknowns, the following relationships are used:

a. Momentum of the system is conserved in the y direction:

$$m_1 V_{1y} + m_2 V_{2y} = m_1 V_{1y}' + m_2 V_{2y}' \tag{3.65}$$

b. Momentum of each body is conserved in the x direction:

$$\begin{aligned} m_1 V_{1x} &= m_1 V_{1x}' \\ m_2 V_{2x} &= m_2 V_{2x}' \end{aligned} \tag{3.66}$$

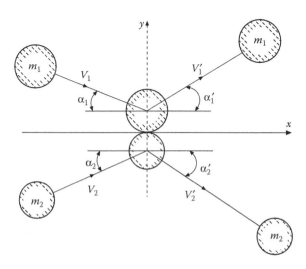

FIGURE 3.31   Impact at an oblique angle.

c. Coefficient of restitution, e, is given by

$$e = \frac{V_1' \sin\alpha_1' + V_2' \sin\alpha_2'}{V_1 \sin\alpha_1 + V_2 \sin\alpha_2} \qquad (3.67)$$

EXAMPLE 3.13

The pendulum (bar plus sphere ball) shown in Figure 3.32a rotates about the z-axis and has a mass of 20 kg (the mass of the bar is 16 kg and the mass of the ball is 4 kg). The pendulum is rotating at $\omega = 10$ rad/s counterclockwise before it strikes the pole when the impulsive force is applied. After the pendulum rebounds, the angular velocity becomes 4 rad/s clockwise. Assume that the pendulum remains in contact with the pole for 0.015 s and that the mass center of the pendulum is 0.67 m from the fixed point O.

a. Determine the corresponding average contact force F between the pendulum and the pole?
b. Determine the pin reaction O_x at the time the pendulum strikes the pole.

SOLUTION

The free-body diagram of the pendulum system is shown in Figure 3.32b.

a. The moment of the linear impulse gives the angular impulse of the impulsive force F. The pendulum has rotation about the fixed point "O". Therefore, taking the moment of the linear impulse about the point "O" gives

$$I\omega_2 = I\omega_1 + \int_{t_1}^{t_2} (-F)L\,dt \qquad (a)$$

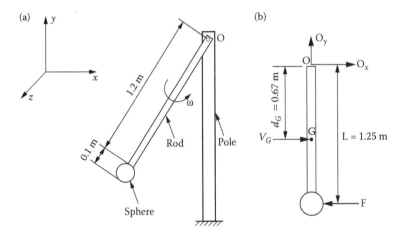

FIGURE 3.32   a) Pendulum system and (b) free-body diagram.

Mass moment of inertia *I* about the z-axis with respect to the fixed point "O" is

$$I\omega_2 = I\omega_1 + \int_{t_1}^{t_2} (-F)L\,dt$$

$$I_{pen} = I_{rod} + I_{sph}$$

$$I_{rod} = \frac{1}{12}m_{rod}l^2 + m_{rod}d^2 = \frac{1}{12}(16)(1.2)^2 + (16)\left(\frac{1.2}{2}\right)^2 = 7.68\ kg\,m^2$$

And

$$I_{sph} = \frac{2}{5}m_{sph}r^2 + m_{sph}d^2 = \frac{2}{5}(4)\left(\frac{0.1}{2}\right)^2 + (4)(1.25)^2 = 6.254\ kg\ m^2$$

Then the total mass moment of inertia of the pendulum is

$$I_{pen} = 7.68 + 6.254 = 13.934\ kg\ m^2$$

Substituting in equation (a)

$$(13.934)(-4) = (13.934)(10) + \left[(-F)L\Delta t\right]$$

$$F = \frac{195.076}{1.25 \times 0.015} = 10,404\ N$$

b. Linear momentum in the x direction is

$$m(V_{G_x})_2 = m(V_{G_x})_1 + \int_{t_1}^{t_2} (O_x - F)\,dt$$

Note that in the above above equation velocity of mass center G is $V_G = \omega r$. Substituting known values

$$(20)\left[(0.67)(-4)\right] = 20\left[(0.67)(10)\right] + (O_x - 10,404)\Delta t$$

$$O_x = \frac{(20)\left[(0.67)(-4)\right] - 20\left[(0.67)(10)\right]}{\Delta t} + 10,404$$

$$O_x = -\frac{187.6}{0.015} + 10,404 = -2102.7\ N$$

Hence,

$$O_x = 2102.7\ N \leftarrow$$

Example 3.14

A steel ball has a velocity of 10 m/s at an angle of 30° as shown in Figure 3.33. Assuming $e = 0.6$ between the steel ball and the steel plate, determine the final velocity and its angle.

SOLUTION

$$e = \frac{V_1' \sin\alpha_1' + V_2' \sin\alpha_2'}{V_1 \sin\alpha_1 + V_2 \sin\alpha_2}$$

Since the steel plate is stationary, V_2 and V_2' are equal to zero; thus, the coefficient of restitution becomes

$$e = \frac{V_1' \sin\alpha_1'}{V_1 \sin\alpha_1} = \frac{V_1' \sin\alpha_1'}{10 \times \sin 30} = 0.6$$

$$V_1' \sin\alpha_1' = (V_1')_y = 0.6 \times 10 \times \sin 30$$

Then

$$(V_1')_y = 3\,\text{m/s}$$

Since no other external forces are applied, the momentum of the ball is conserved in the x-direction:

$$m_1(V_1)_x = m_1(V_1')_x$$

or

$$(V_1)_x = (V_1')_x \quad \Rightarrow \quad (V_1')_x = 10 \times \cos 30 = 8.66\,\text{m/s}$$

Substituting velocity components yields

$$V_1' = \sqrt{(V_1')_x^2 + (V_1')_y^2} = \sqrt{8.66^2 + 3^2} = 9.16\,\text{m/s}$$

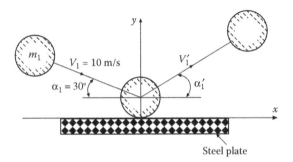

FIGURE 3.33 Impact between the steel ball and the steel plate.

But

$$\tan\alpha_1' = \frac{(V_1')_y}{(V_1')_x} = \frac{3}{8.66} = 0.3464$$

and

$$\alpha_1' = 19.1°$$

PROBLEMS

3.1. Consider a pulley system carrying two loads by inextensible wires as shown in the figure below, adapted from Singer 1975. If the initial angular velocity of pulley B is 8 rad/s counterclockwise and D is decelerating at a constant rate of 2 ft s², determine the distance that weight A travels before coming to rest.

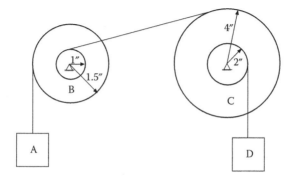

3.2. As shown in the following figure, wheel A rotates counterclockwise at an angular velocity, ω_A, of 250 rpm. If the angular velocity of the arm AB is $\omega_{AB} = 60$ rpm clockwise, determine the angular velocity of wheel B.

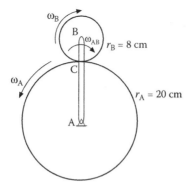

3.3. As shown in the following figure, a rigid bar is connected by pins to two slider blocks at points A and B. At any instant time, the velocity of

A is $V_A = 8$ m/s and acceleration is $a = 3$ m/s^2. Find the angular velocity and acceleration of the rigid bar AB.

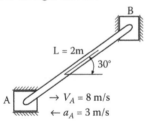

3.4. As shown in the following figure, a 2-lb force is applied on a solid disk
at point A. Determine the angular acceleration of the disk, assuming
that the solid disk weighs 10 lb and the radius of gyration about the
center of mass G is 2 in.

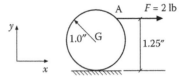

3.5. The gear shown in the following figure rolls on the stationary lower
rack. The velocity of the gear at the center of mass G is 1.5 m/s with
the direction toward right. Determine (a) the angular velocity of the
gear, and (b) the velocity of the upper rack R.

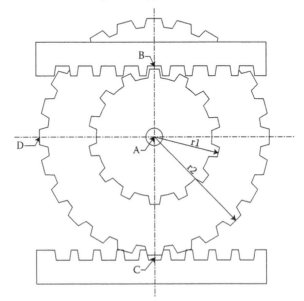

3.6.   Using the following figure, determine the acceleration of each block in
       Example 3.8.

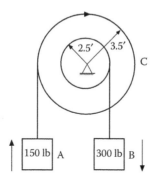

3.7.   A pendulum with a spherical ball attached is supported by a pin
       joint at point A as shown in the following figure. The weights of
       the rod and sphere are 20 lb and 40 lb, respectively. Determine the
       horizontal component of the reaction force at point A when the
       pendulum swings due to a force of 50 lb at a distance of 3 ft from
       point A.

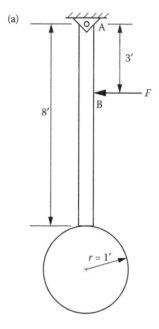

(*Continued*)

(b)

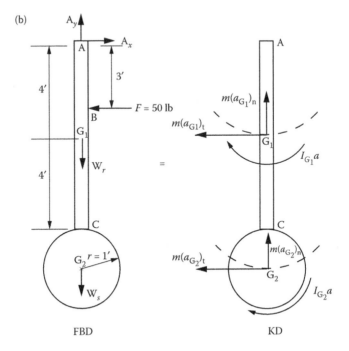

FBD   KD

3.8.   Determine the torque that should be applied to the disk shown in the following figure when the velocity of the 2-kg mass reaches 6 m/s in 6 s from rest. Assume that the coefficient of kinetic friction, μ_k, between the mass and the plane is 0.2.

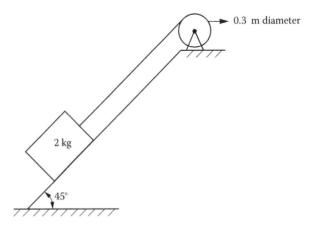

0.3 m diameter

2 kg

45°

3.9.   As shown in the following figure, a 4-m-long steel I beam weighing 5 kg is supported by a frictionless pin at one end and a guy wire at the other end. The beam's radius of gyration about an axis perpendicular

to G is 0.7 m. Determine the instantaneous reactions at support A
when the wire suddenly snaps.

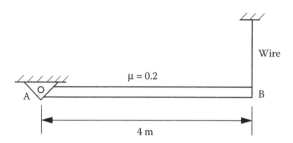

3.10. As shown in the following figure, a cord is wrapped around the inner
radius of a stepped solid disk. If the cord is pulled with a constant tension
of 40 lb when the disk is initially at rest, determine the angular velocity
of the cylinder when 12 ft of cord has been pulled from the disk. Assume
that the weight of the disk is 350 lb and the radius of gyration is 1.5 ft.

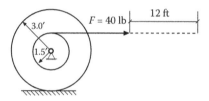

3.11. The solid cylinder shown in the following figure is 3 ft in diameter and
weighs 500 lb. A 150-lb force in the y-direction is applied by a cord
wrapped around it. Determine the coefficient of friction to prevent
slipping.

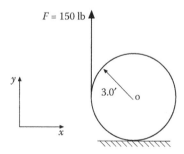

3.12. As shown in the following figure, a 0.015-kg bullet is fired horizon-
tally into a 2-kg block. The bullet embeds itself in the block and they
both move a distance of 0.5 m to the right until brought to a stop by
the action of the linear spring (with $k = 450$ N/m) and the friction with

the floor. If the spring is initially uncompressed, determine the speed of the bullet V_1 to move the block a distance of 0.5 m. Assume that the coefficient of friction between the floor and the block is 0.2.

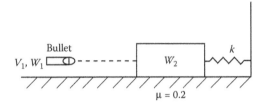

3.13. As shown in the following figure, the 150-lb sand box is initially at rest on the 45° inclined plane. A 1-lb projectile is fired into the sand box at a velocity of 1100 ft/s parallel to the plane. Determine how far the box and the projectile move uphill.

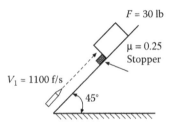

3.14. As shown in the following figure, a 0.015-kg bullet is fired with a velocity of 800 m/s into a 0.5-kg sand box mounted on wheels. After impact the box strikes a spring and is brought to rest at a distance of 0.15 m. Determine the constant of the spring.

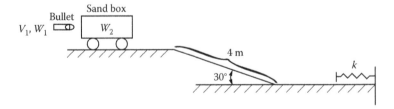

3.15. If the initial velocity of ball A, shown in the following figure, is 4 m/s and the coefficient of restitution, e, is 0.3, determine the velocity of the ball after it strikes a 0.5-kg plate. Also determine how far the spring will be compressed after the ball strikes the plate. Assume that the spring is initially at its free length and ignore gravitational effects.

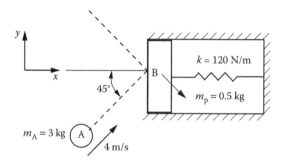

BIBLIOGRAPHY

Housner, G. W. and Hudson, E. D. *Applied Mechanics: Dynamics*, D. Van Nostrand Company, Inc, New York, 1959.

Meriam, J. L. *Engineering Mechanics, Volume 2: Dynamics*, John Wiley & Sons, New York, 1980.

Riley, F. W. and Sturgis, D. L. *Engineering Mechanics. Mechanics: Dynamics*, John Wiley & Sons, New York, 1996.

Shelley, J. F. *Engineering Mechanics: Mechanics and Dynamics*, McGraw-Hill, New York, 1980.

Singer, F. L. *Engineering Mechanics, Part II: Dynamics*, Harper & Row, New York, 1975.

4 Solid Mechanics

4.1 INTRODUCTION

The science of the strength of materials developed after the development of the principle of statics. In statics, applied forces and the equilibrium of force systems acting on structural members have been studied. Dimensional changes and failure of these structural members under applied loading conditions have not been examined.

The study of the strength of materials is a branch of engineering that deals with the relationship between the loads applied to solid elastic bodies and the internal stresses causing deformation of the bodies. In general, studying the strength of materials provides a more comprehensive explanation of the behavior of solids under applied loads. This chapter contains brief introductory information related to the strength of materials and is, for the most part, restricted to a study of relatively simple structures and machines. However, a good understanding of basic facts can be used to solve more complex design or analysis problems.

4.2 STRESS ANALYSIS

4.2.1 UNIFORM NORMAL STRESS AND STRAIN

4.2.1.1 Uniform Normal Stress

Consider an axially loaded beam as shown in Figure 4.1. If the tensile load applied on the infinitesimal area dA is dp and is assumed to be uniform over the cross section of the beam shown in Figure 4.1, the normal stress at any infinitesimal area is defined by

$$\sigma = \frac{dp}{dA} \text{ (psi or MPa)} \tag{4.1}$$

or over the entire area as

$$\sigma = \frac{p}{A} \tag{4.2}$$

The stress given by Equation 4.2 is called the uniform normal stress. For the normal stress to be uniform over the entire cross-sectional area, the following requirements must be satisfied:

1. The beam member must be straight.
2. The beam member must have constant cross-sectional area.

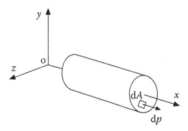

FIGURE 4.1   Axially loaded beam.

3. The beam material must be homogeneous; that is, every point of the beam must have the same material properties.
4. The applied load must be along the centroidal axis of the beam cross section.

4.2.1.2   Uniform Normal Strain

The axial loading shown in Figure 4.1 occurs frequently in structural members and in mechanical component design problems. To simulate this type of loading in the laboratory, a standard test specimen as shown in Figure 4.2 is used. Laboratory tests applying tension or compression are in general performed at a constant slow rate of deformation using tension–compression testing machines. The tension test is usually performed by using ductile material (fails by yielding) such as steel, aluminum, plastic, and so on, while the compression test is often performed using brittle materials (fails by rupture) such as concrete, cast iron, glass, and so on.

Assume that a test specimen made of steel is pulled by a tension–compression testing machine. As tensile forces are slowly applied to the end of the specimen at a constant rate of deformation, the stress–strain curve shown in Figure 4.3 can be obtained.

Consider the one-dimensional thin bar of length L shown in Figure 4.4, subjected to applied load P. The amount of stretch or elongation is called strain. The elongation per unit length is called unit strain and is dimensionless. If the strain of a bar along length L is uniform, it is called uniform normal strain.

$$\varepsilon = \frac{\delta}{L} \text{ in./in.} \tag{4.3}$$

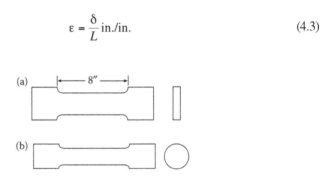

FIGURE 4.2   Test specimens (a) rectangular cross section and (b) circular cross section.

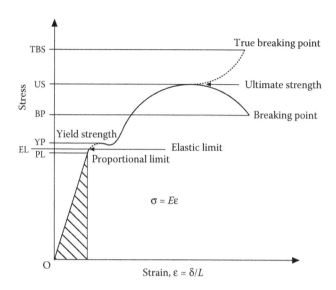

FIGURE 4.3 Stress–strain curve.

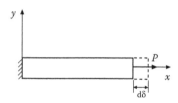

FIGURE 4.4 Axially loaded thin bar.

Some of the definitions shown in Figure 4.3 are listed below.

1. *Proportional Limit (PL)*: As seen in Figure 4.3, the line segment from origin O to point PL is straight. In other words, between these two points stress is a linear function of strain. The proportional limit (PL) is defined by the highest stress point of the straight line. This linear relationship was first noticed by Robert Hooke in 1678 and is known as Hooke's law.
2. *Elastic Limit*: The highest stress level that a material is capable of withstanding without permanent deformation when the applied load is removed is known as the elastic limit.
3. *Yield Point*: The point at which there is an increase in strain with no corresponding increase in stress is known as the yield point of the material. The yield point represents the transition from elastic behavior to plastic behavior.
4. *Ultimate or Tensile Strength* (S_{ut}): The highest tensile stress that a material can withstand before rupture is referred to as the ultimate or tensile strength.

5. *Modulus of Elasticity* (E): The slope of the straight line portion of the stress–strain curve is called modulus of elasticity or Young's modulus and is given as

$$E = \frac{\Delta\sigma}{\Delta\varepsilon} = \frac{\sigma}{\varepsilon} \qquad (4.4)$$

where σ is any stress on the straight line and ε is the strain corresponding to σ. Young's modulus, E, has the same units as stress (i.e., psi) and its value for various materials is tabulated in material handbooks. For example, E is approximately 30×10^6 psi for steel. Using Equations 4.2 and 4.3, Equation 4.4 can be rearranged as

$$\delta = \frac{PL}{AE} = \frac{\sigma L}{E} \qquad (4.5)$$

6. *Poisson's Ratio*: Consider the two-dimensional bar shown in Figure 4.5. When a bar is subjected to a tensile load P, there is not only an increase in the length of the bar in the direction of the applied load but also a decrease in the lateral dimensions perpendicular to the load. The ratio of the strain in the lateral direction to that in the axial direction is called Poisson's ratio and can be given as

$$\mu = \frac{\text{Lateral strain}}{\text{Axial strain}} = -\frac{\varepsilon_1}{\varepsilon_a} \qquad (4.6)$$

where

$$\varepsilon_1 = -\frac{\delta y}{y} \text{ (contraction)}$$
$$\varepsilon_a = \frac{\delta x}{x} \text{ (elongation)} \qquad (4.7)$$

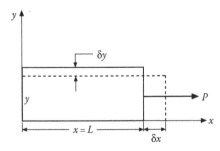

FIGURE 4.5   Two-dimensional bar subjected to a tensile load.

Poisson's ratio, μ, is constant for any homogeneous isotropic material. For elastic material within the elastic range, the value of Poisson's ratio varies between 0.25 and 0.30. Beyond the elastic limit of the material, this value is often assumed to be 0.5.

4.2.2 UNIFORM SHEAR STRESS AND STRAIN

So far, normal stresses produced by loads that are perpendicular to the area upon which they act have been discussed. Another category of stress is shear stress, which is produced by loads that are parallel to the area upon which they act. Figure 4.6a shows a rectangular element with shear force V acting on it. Figure 4.6b shows shear stress τ and shear strain γ produced by shear force V.

The uniform shear stress can be determined as

$$\tau = \frac{V}{A} \tag{4.8}$$

where V is the shear force and A is the area over which it acts.

The relationship between shear stress and shear strain is called the shear modulus, G, and is given as

$$G = \frac{\tau}{\gamma} \tag{4.9}$$

The three elastic constants, μ, E, and G, are related to each other as

$$E = 2G(1 + \mu) \tag{4.10}$$

4.2.3 THERMAL STRESS AND STRAIN

Temperature change causes strain in a bar and, consequently, it shrinks or expands. This deformation is given as

$$\delta_T = \alpha L(\Delta T) \tag{4.11}$$

where α is the coefficient of thermal expansion, ΔT is the temperature change, and L is the length of the bar. Usually, expansion joints are used to eliminate the stress in a structure caused by temperature change, also known as thermal stress.

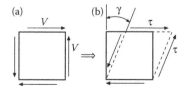

FIGURE 4.6 Shear stress and shear strain produced by shear force: (a) undeformed shape and (b) deformal shape.

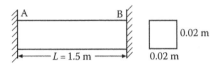

FIGURE 4.7   Rod between two steel walls.

Example 4.1

As shown in Figure 4.7, a 1.5-m rod is welded between two steel walls. At a room temperature of 20°C, the thermal stress on the rod is approximately zero. Determine the thermal stress on the rod when the temperature is reduced to −20°C. Assume that the walls are rigid, $\alpha = 11.9 \times 10^{-6}/°C$, and $E = 200$ GPa.

SOLUTION

Due to the drop in temperature the rod will tend to shrink, but since the walls are rigid the rod will be elongated under tension. If the rod were free to shrink, deformation would be

$$\delta_T = \alpha L (\Delta T)$$

and

$$\delta = \frac{PL}{AE}$$

substituting which yields

$$\alpha L(\Delta T) = \frac{PL}{AE} = \frac{\sigma L}{E}$$

Solving for σ from the above equation, we have

$$\sigma = E\alpha\,\Delta T = (200 \times 10^9)(11.9 \times 10^{-6})[20 - (-20)] = 95.2 \times 10^6\,\text{Pa}$$

Note that the length of the bar does not affect the stress due to change in temperature.

4.2.4   Normal Bending Stress in Beams

As shown in Figure 4.8, the bending moment M applied on a beam having a longitudinal plane of symmetry produces a normal stress σ at distance y from the neutral axis of the beam. This stress is given as

$$\sigma = \frac{My}{I} \qquad (4.12)$$

If the maximum bending stress is desired, substitute $y = c$ to give

$$\sigma_{max} = \frac{Mc}{I} \qquad (4.13)$$

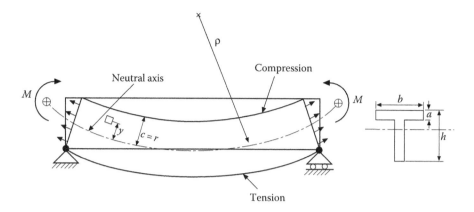

FIGURE 4.8 Normal stresses in the beam.

where c is the distance from the neutral axis to the outermost fiber. For example, in the case of circular cross section $c = r$. As seen from Figure 4.8, since $y = 0$ at the neutral axis the bending stress is zero. The relationship between the bending moment and the radius of curvature ρ is given as

$$\frac{1}{\rho} = \frac{M}{EI} \tag{4.14}$$

where

$$\frac{1}{\rho} = \frac{d^2y}{dx^2} \tag{4.15}$$

Equations 4.14 and 4.15 will be used in beam deflection analysis.

Example 4.2

Determine the maximum normal stress due to bending and its location for the beam shown in Figure 4.9.

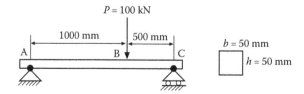

FIGURE 4.9 Beam under bending stress.

SOLUTION

To determine the reaction forces, draw an FBD (see Figure 4.10) and apply the equilibrium equations

$$\sum F_y = 0 \quad \Rightarrow \quad -100 + R_{Ay} + R_{Cy} = 0$$

$$R_{Ay} + R_{Cy} = 100 \text{ lb}$$

$$\sum M_A = 0 \quad \text{(assuming counterclockwise moments are positive)}$$

$$-1000 \times 1 + R_{Cy} \times 1.5 = 0$$

$$R_{Cy} = \frac{10,000 \times 1}{1.5} = 66.7 \text{ kN}$$

Then

$$R_{Ay} = 100 - R_{Cy} = 100 - 66.7 = 33.3 \text{ kN}$$

The bending moment is maximum where the shear force is zero. Zero shear force occurs at the point where the load P is applied. Thus, the maximum bending moment is $M_{max} = 33.3 \times 10^3$ kN mm. The bending moment at any section x is equal to the area of the shear diagram to the left or right of that section. For example, the maximum bending moment at $x = 1$ m is

$$M_{x=14''} = \int_{x=0}^{x=1000} V \, dx = V \int_{x=0}^{x=1000} dx = (33.3)x \big|_0^{1000}$$

$$M_{x=1000} = 33.3 \times 10^3 \text{ kNmm}$$

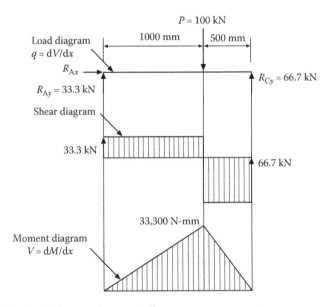

FIGURE 4.10 Load, shear, and moment diagram.

Then the maximum bending stress is

$$\sigma_{max} = \frac{M_{max}c}{I}\sqrt{2}$$

$$c = \frac{h}{2} = \frac{50}{2} = 25 \text{ mm}$$

$$I = \frac{bh^3}{12} = \frac{25 \times 25^3}{12} = 32,552 \text{ mm}^4$$

$$\sigma_{max} = \frac{33.3 \times 10^3 \times 25}{32,552} = 25.6 \text{ N/mm}^2$$

The following statements can be made to explain the relationship between load, shear, and bending moment diagrams.

1. The load at any section is equal to the derivative of the shear diagram at the same section,

$$q = \frac{dV}{dx} \tag{4.16}$$

   where q is the load.
2. The shear at any section is equal to the derivative of the bending moment diagram at the same section,

$$V = \frac{dM}{dx} \tag{4.17}$$

4.2.5 SHEAR STRESS IN BEAMS

If the shear stress is uniform over the cross section, the uniform shear stress can be calculated as

$$\tau = \frac{V}{A} \tag{4.18}$$

where V is the shear force in the beam and A is the cross-sectional area on which V acts. However, shear stress cannot be assumed to be uniform in a beam.

Consider a beam composed of fibers as shown in Figure 4.11. A load acting on the beam will cause the fiber layers to slip on each other (Figure 4.11b). This creates a horizontal shear in the beam which is defined by

$$\tau = \frac{V\bar{Q}}{Ib} \tag{4.19}$$

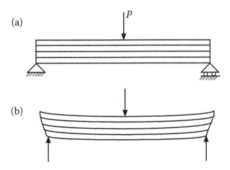

FIGURE 4.11 Beam composed of fibers (a) undeformed shape and (b) deformed shape.

where V is the transverse shear force, I is the area moment of inertia about the neutral axis, b is the width of the beam, and $\bar{Q}$ is the first moment of the cross-sectional area on which the shear stress occurs.

$$\bar{Q} = \int y \, dA = \bar{y}\bar{A} \qquad (4.20)$$

where $\bar{A}$ is the shear area above the neutral axis and $\bar{y}$ defines the location of the centroid of the shear area with respect to the neutral axis. Since V and I in Equation 4.19 are constants, the maximum shear stress can be obtained when the quantity $\bar{y}\bar{A}$ is maximum.

Example 4.3

Determine the maximum shear stress for a rectangular shape, as shown in Figure 4.12.

SOLUTION

Shear stress is maximum at the neutral axis. Therefore, area $\bar{A}$ is the area above the neutral axis.

$$\bar{A} = \frac{h}{2} \times b$$

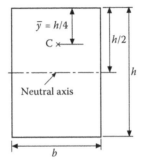

FIGURE 4.12 Maximum shear stress for a rectangular shape.

and

$$I = \frac{bh^3}{12}$$

If the centroid of the area above the neutral axis $\bar{y} = h/4$, then the maximum shear stress is

$$\tau_{max} = \frac{V}{Ib}\bar{A}\bar{y} = \frac{V}{(bh^3/12) \times b}\left(\frac{h}{2} \times b\right)\left(\frac{h}{4}\right)$$

Therefore,

$$\tau_{max} = \frac{3V}{2A}$$

Example 4.4

Draw an FBD of the load, shear, and moment diagrams of the beam shown in Figure 4.13. Determine the maximum bending moment and shear stress at $x = 45''$.

SOLUTION

Draw an FBD as shown in Figure 4.14 and apply the following equilibrium equations to determine the reaction forces:

$$\sum F_y = 0$$
$$R_{Ay} + R_{Cy} = 20 \times 30 + 200$$
$$R_{Ay} + R_{Cy} = 800 \text{ lb}$$

$$\sum M_A = 0$$
$$-600 \times 15 + 60R_{Cy} - 80 \times 200 = 0$$
$$R_{Cy} = 417 \text{ lb}$$

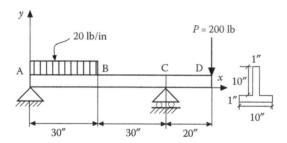

FIGURE 4.13 Beam under loading stress.

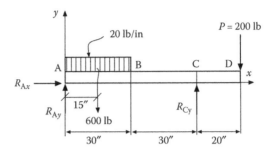

FIGURE 4.14   Free-body diagram of beam.

Therefore,

$$R_{Ay} = 383 \text{ lb}$$

$$\Sigma_{Fx} = 0$$

Then

$$R_{Ax} = 0$$

Since shear stress is equal to zero at two points, $x_1 = 19.15''$ and $x_1 = 60''$, there are two possible maximum bending moments (see Figure 4.15). Location of x_1 can be calculated by $383 = 20x_1 \Rightarrow x_1 = 19.5''$ (20 lb/in. at x should balance the load of 383 lb as shown in the shear diagram). The maximum positive moment occurs at $x_1 = 19.15''$. Then

$$
\begin{aligned}
M_{x=19,15''} &= \text{Area of triangle of shear diagram} \\
&= \frac{383 \times 19.15}{2} = 3667 \text{ lb-in.}
\end{aligned}
$$

The maximum negative moment occurs at $x = 60''$

$$
\begin{aligned}
M_{x=60''} &= \text{Area of rectangle of shear diagram} \\
&= 200 \times 20 = 4000 \text{ lb-in.}
\end{aligned}
$$

The neutral axis $\bar{y}$ can be located by (Figure 4.16)

$$\bar{y} = \frac{\sum y_i A_i}{\sum A_i} = \frac{(6)(10)(1) + (0.5)(10)(1)}{(10)(1) + (10)(1)} = 3.25''$$

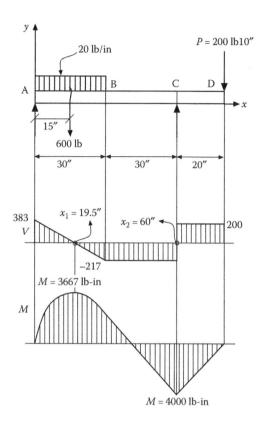

FIGURE 4.15 Load, shear, and bending moment diagram.

The moment of inertia I about the neutral axis is

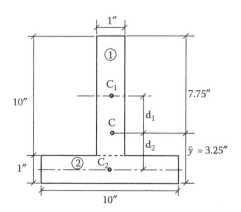

FIGURE 4.16 Locating the neutral axis.

$$I = I_1 + I_2$$

$$I_1 = \frac{bh^3}{12} + A_1 d_1^2 = \frac{1 \times 10^3}{12} + (1 \times 10)(2.75)^2 = 159 \text{ in.}^4$$

$$I_2 = \frac{bh^3}{12} + A_2 d_2^2 = \frac{10 \times 1^3}{12} + (1 \times 10)(3.25 - 0.5)^2 = 76 \text{ in.}^4$$

$$I = I_1 + I_2 = 159 + 76 = 235 \text{ in.}^4$$

The maximum compressive stress created by positive $M = 3667$ lb-in. at the top fiber of the beam is

$$\sigma_{max} = \frac{Mc}{I} = \frac{3667 \times 7.75''}{235} = 120.9 \text{ psi} \quad (\text{compression})$$

The maximum tensile stress created by negative $M = 4000$ lb-in. at the bottom fiber of the beam is

$$\sigma_{max} = \frac{Mc}{I} = \frac{4000 \times 7.75''}{235} = 131.9 \text{ psi} \quad (\text{tension})$$

From Figure 4.17, shear stress at $x = 45''$ is

$$383 - V - 600 = 0 \quad \Rightarrow \quad V = 217 \text{ lb}$$

From Figure 4.16, $\bar{Q}$ is

$$\bar{Q} = \bar{A}\bar{y} = (7.75 \times 1'')\left(\frac{7.75}{2}\right) \quad \Rightarrow \quad \bar{Q} = 30 \text{ in.}^3$$

Then the maximum shear is

$$\tau_{max} = \frac{V\bar{Q}}{Ib} = \frac{(217)(30)}{76 \times 1''} = 85.55 \text{ psi}$$

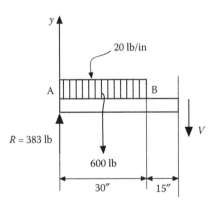

FIGURE 4.17   Free-body diagram for shear stress at $x = 45''$.

4.2.6 STRESS IN THIN-WALLED PRESSURE VESSELS

Internal pressure produces tensile stresses in the wall of pressure vessels. The stresses in the wall of the vessels vary over the thickness of the vessel wall. Stress is maximum at the inner surface and minimum at the outer surface. In a thin-walled pressure vessel the thickness of the wall of the vessel is less than the principal radius of curvature and the stresses are assumed to be uniformly distributed over the thickness of the wall.

Figure 4.18 shows a pressure vessel subjected to internal pressure p. As shown in the figure, internal pressure produces hoop or circumferential stress σ_h and axial or longitudinal stress σ_a. Applying equilibrium equation to Figure 4.18b,

$$\sum F_n = 0 \tag{4.21}$$

$$\sigma_h(2Lt) - p(DL) = 0 \tag{4.22}$$

$$\sigma_h = \frac{pD}{2t} \quad \text{(hoop stress)} \tag{4.23}$$

Applying equilibrium equation to Figure 4.18c,

$$\sum F_a = 0 \tag{4.24}$$

$$\sigma_a(\pi Dt) - p\frac{\pi D^2}{4} = 0 \tag{4.25}$$

$$\sigma_a = \frac{pD}{4t} \quad \text{(axial stress)} \tag{4.26}$$

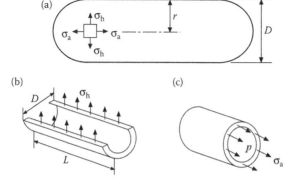

FIGURE 4.18 (a) Thin-walled pressure vessel, (b) free-body diagram showing hoop stress, and (c) free-body diagram showing axial stress.

From Equations 4.23 and 4.26, it can be noted that the hoop stress is twice the axial stress.

Example 4.5

The hoop and axial stresses of a pressure vessel are $\sigma_h = 330 \times 10^6$ N/m^2 and $\sigma_a = 165 \times 10^6$ N/m^2, respectively. Determine the pressure in the vessel. Assume $t = 0.6$ mm and $D = 600$ mm.

SOLUTION

Either Equation 4.23 or Equation 4.26 can be used to determine the internal pressure

$$p = \frac{2t\sigma_h}{D} = \frac{2 \times 6 \times 10^{-4} \times 330 \times 10^6}{600 \times 10^{-3}} = 6.6 \times 10^5 \, \text{N/m}^2$$

4.2.7 COMBINED STRESS

Stress analysis discussed so far is with regard to structural members with only one type of load. Often, combinations of these loads (axial, bending, and torsion) are applied on members. The basic stress formulas with respect to elastic bars and cylindrical pressure vessels are as follows:

a. Elastic Bars:
   Tension or compression: $\sigma = \pm P/A$
   Torsion: $\tau = Tr/J$ (circular section)
   Bending: $\sigma = \pm Mc/I$
   Shear: $\tau = V\bar{Q}/Ib$
b. Cylindrical Pressure Vessels:
   Hoop stress: $\sigma_h = pD/2t$
   Axial stress: $\sigma_a = pD/4t$

When a structural member is subjected to two or more types of load simultaneously, stresses at a point on the surface of the member consists of both normal and shear components. If the material is linearly elastic, stresses resulting from combined loads can be determined by the principle of superposition.

Normal stresses due to axial and bending forces can be directly summed, as also in the case of shear stresses, if they act in the same direction. Although normal and shear forces cannot be summed directly, they can be combined using appropriate failure theories and stress–strain analysis.

4.3 STRESS–STRAIN ANALYSIS

A general three-dimensional stress element is shown in Figure 4.19. The stresses acting upon an element define the state of stress at that point in the body. Stresses acting in the directions of x, y, and z are σ_x, σ_y, and σ_z, respectively, and are denoted

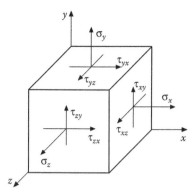

FIGURE 4.19 Three-dimensional stress element.

as normal stresses. Shear stresses occur in pairs of equal magnitude on the orthogonal faces of an element. The first subscript of shear stress denotes the coordinate normal to the element face and the second subscript indicates the direction of the stress. For example, τ_{xy} is the shear stress acting upon the x face of the element in the y-direction.

Since the shear stresses acting on the element are equal in magnitude ($\tau_{yx} = \tau_{xy}$, $\tau_{zx} = \tau_{xz}$, $\tau_{yz} = \tau_{zy}$) six components of stress (σ_x, σ_y, σ_z, τ_{xy}, τ_{xz}, τ_{yz}) are required to completely define the state of stresses at a point in a loaded body.

4.3.1 Plane Stress

As shown in Figure 4.20, the state of plane stress in the x–y plane is defined by $\sigma_z = \tau_{xz} = \tau_{yz} = 0$. Shear stress components will be specified by clockwise (CW), considered positive, and counterclockwise (CCW), considered negative, properties. Similarly, normal stress components will be specified by tension (T), considered positive, and compression (C), considered negative, properties.

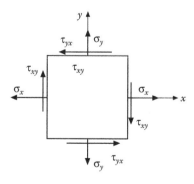

FIGURE 4.20 Two-dimensional stress element.

4.3.2 STRESS ON AN INCLINED PLANE

Often, it is desirable to define the state of stress on a plane inclined at an angle as shown in Figure 4.21.

The normal and shear stresses, σ_n and τ_t, respectively, acting on an inclined plane can be determined from the equilibrium of the element shown in Figure 4.21.

$$\sigma_n = \frac{\sigma_x + \sigma_y}{2} - \frac{\sigma_x - \sigma_y}{2} \cos 2\alpha + \tau_{xy} \sin 2\alpha \tag{4.27}$$

$$\tau_t = \left(\frac{\sigma_x - \sigma_y}{2} \right) \sin 2\alpha + \tau_{xy} \cos 2\alpha \tag{4.28}$$

To define a coordinate system which has zero shear stress, set Equation 4.28 equal to zero.

$$\left(\frac{\sigma_x - \sigma_y}{2} \right) \sin 2\alpha + \tau_{xy} \cos 2\alpha = 0 \tag{4.29}$$

or

$$\tan 2\alpha = \frac{2\tau_{xy}}{\sigma_y - \sigma_x} \tag{4.30}$$

In Equation 4.30, α defines the inclined plane which has normal stress but zero shear stress. These stresses are termed principal (maximum and minimum) stresses and are denoted by σ_1 and σ_2. They are given as

$$\sigma_{1,2} = \frac{\sigma_x + \sigma_y}{2} \pm \sqrt{\left(\frac{\sigma_x - \sigma_y}{2} \right)^2 + \tau_{xy}^2} \tag{4.31}$$

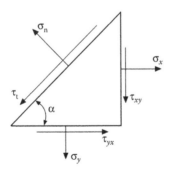

FIGURE 4.21   Stresses on an inclined plane.

Similarly, for certain values of angle α, the maximum shear stress $\tau_{1,2}$ can be defined as

$$\tau_{1,2} = \mp \sqrt{\left(\frac{\sigma_x - \sigma_y}{2}\right)^2 + \tau_{xy}^2} \qquad (4.32)$$

4.3.3 Mohr's Circle of Stress

The graphical representation of handling stress transformation and principal stress problems is called Mohr's circle. Mohr's circle, as shown in Figure 4.22, for a given element can be drawn using the following steps:

1. Lay off normal stress, σ, and shear stress, τ, axes.
2. Locate the center of Mohr's circle along the σ axis by calculating

$$a = \frac{\sigma_x + \sigma_y}{2} \qquad (4.33)$$

3. Locate point A by using σ_x (+) and τ_{xy} (CCW, −) and locate point B by using σ_y (+) and τ_{xy} (CW, +).
4. Connect points A and B, passing through center C, with a straight line.

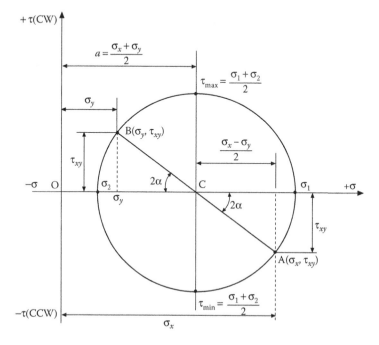

FIGURE 4.22 Mohr's circle.

5. Draw a circle with the radius AC.
6. The circle intersects σ axis at σ_1 (maximum principal stress) and σ_2 (minimum principal stress).

The radius of Mohr's circle is given by

$$\tau_{1,2} = r = \sqrt{\left(\frac{\sigma_x - \sigma_y}{2}\right)^2 + \tau_{xy}^2} = \frac{\sigma_1 - \sigma_2}{2} \tag{4.34}$$

From Mohr's circle, angle 2α can easily be found as

$$\tan 2\alpha = \frac{\tau_{xy}}{(\sigma_x - \sigma_y)/2} \tag{4.35}$$

Example 4.6

The stress element shown in Figure 4.23 has $\sigma_x = 80$ MPa, $\sigma_y = 100$ MPa, and $\tau_{xy} = 50$ MPa. Determine the principal stresses and directions and show these on a stress element correctly. Determine the corresponding normal stresses with respect to the x-axis on the original element shown in Figure 4.23.

SOLUTION

Locate the centroid

$$a = OC = \frac{\sigma_x + \sigma_y}{2} = \frac{80 + 100}{2} = 90 \text{ MPa}$$

$$r = \sqrt{\left(\frac{\sigma_x - \sigma_y}{2}\right)^2 + \tau_{xy}^2} = \sqrt{\left(\frac{80 - 100}{2}\right)^2 + 50^2} = 51 \text{ MPa}$$

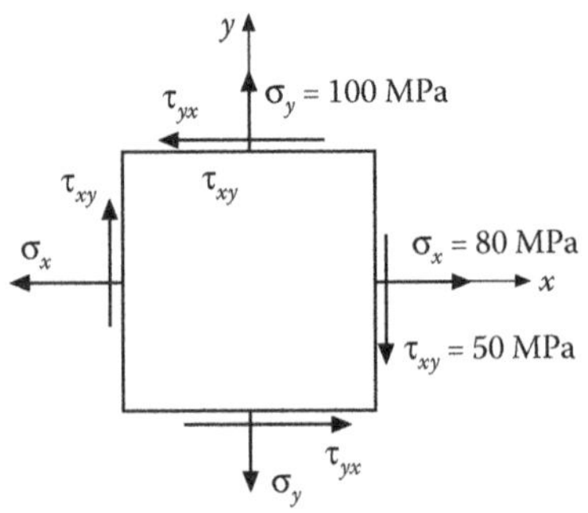

FIGURE 4.23 Stress element.

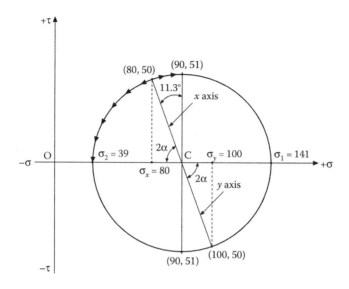

FIGURE 4.24 Mohr's circle.

Therefore,

$$\tau_{max,min} = \mp r = \mp 51\,\text{MPa}$$

Since

$$\sigma_{1,2} = a \mp r$$
$$\sigma_1 = 90 + 51 = 141\,\text{MPa}$$
$$\sigma_2 = 90 - 51 = 39\,\text{MPa}$$

from Figure 4.24

$$\tan 2\alpha = \frac{50}{90 - 80} \Rightarrow 2\alpha = 78.7°$$

on Mohr's circle (see Figure 4.24) and $\alpha = 39.35°$ on the original element (see Figure 4.25). Note that angles on Mohr's circle are twice those on a given element. Stresses with respect to the original x-axis are shown in Figure 4.25.

4.3.4 Principal Stresses in Three Dimensions

If the values σ_x, σ_y, σ_z, τ_{xy}, τ_{xz}, τ_{yz}, shown in Figure 4.26 are given the three roots of the following equation, will provide the principal stresses.

$$\begin{vmatrix} \sigma - \sigma_x & -\tau_{xy} & -\tau_{zx} \\ -\tau_{xy} & \sigma - \sigma_y & -\tau_{yz} \\ -\tau_{zx} & -\tau_{yz} & \sigma - \sigma_z \end{vmatrix} = 0 \tag{4.36}$$

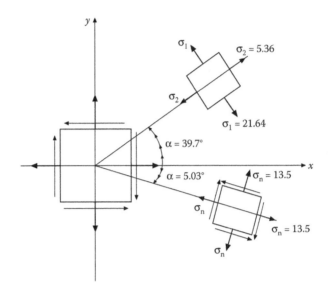

FIGURE 4.25   Stresses with respect to the original x axis.

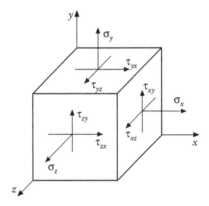

FIGURE 4.26   Stresses in three dimensions.

or

$$
\sigma^3 - \left(\sigma_x + \sigma_y + \sigma_z\right)\sigma^2
$$
$$
+ \left(\sigma_x\sigma_y + \sigma_y\sigma_z + \sigma_z\sigma_x - \tau_{xy}^2 - \tau_{xz}^2 - \tau_{zx}^2\right)\sigma \qquad (4.37)
$$
$$
- \left(\sigma_x\sigma_y\sigma_z + 2\tau_{xy}\tau_{yz}\tau_{zx} - \sigma_x\tau_{yz}^2 - \sigma_y\tau_{zx}^2 - \sigma_z\tau_{xy}^2\right) = 0
$$

After the three roots, σ_1, σ_2, and σ_3 are determined, Mohr's circle for three dimensions, shown in Figure 4.27, can be drawn. To determine the direction

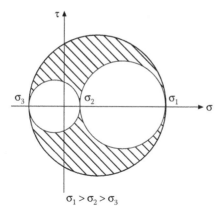

$$\sigma_1 > \sigma_2 > \sigma_3$$

FIGURE 4.27   Mohr's circle in three dimensions.

cosines l, m, and n of principal stresses, the following three simultaneous equations are used:

$$(\sigma - \sigma_x)l - \tau_{xy}m - \tau_{zx}n = 0$$
$$-\tau_{xy}l + (\sigma_1 - \sigma_y)m - \tau_{yz}n = 0 \qquad (4.38)$$
$$-\tau_{zx}l - \tau_{yz}m + (\sigma_1 - \sigma_z)n = 0$$

where

$$l^2 + m^2 + n^2 = 1 \qquad (4.39)$$

and the direction cosines are:

- $l = \cos\theta$, where θ is the angle of direction cosines of principal stress with respect to the x-axis.
- $m = \cos\varphi$, where φ is the angle of direction cosines of principal stress with respect to the y-axis.
- $n = \cos\psi$, where ψ is the angle of direction cosines of principal stress with respect to the z-axis.

Equations 4.38 and 4.39 can be rewritten to calculate three principal stresses, σ_1, σ_2, and σ_3.

Example 4.7

Assume that an element is subjected to the following stress field:

$$\sigma_x = 8000\,\mathrm{psi}\ (T), \qquad \tau_{xy} = 3000\,\mathrm{psi}$$
$$\sigma_y = -4000\,\mathrm{psi}\ (C), \qquad \tau_{yz} = 3000\,\mathrm{psi}$$
$$\sigma_z = -4000\,\mathrm{psi}\ (C), \qquad \tau_{zx} = 0$$

Determine the principal stresses σ_1, σ_2, and σ_3 and calculate the direction cosine of σ_1.

SOLUTION

Substituting the given values into Equation 4.37 yields the following principal stresses:

$$\sigma_1 = 8.844 \times 10^3 \, \text{psi(T)}$$
$$\sigma_2 = -7.131 \times 10^3 \, \text{psi(C)}$$
$$\sigma_3 = -1.713 \, \text{psi(C)}$$

Rewriting Equation 4.38 to calculate the direction cosine of σ_1 gives

$$(\sigma_1 - \sigma_x)l_1 - \tau_{xy}m_1 - \tau_{zx}n_1 = 0$$
$$-\tau_{xy}l_1 + (\sigma_1 - \sigma_y)m_1 - \tau_{yz}n_1 = 0$$
$$-\tau_{zx}l_1 - \tau_{yz}m_1 + (\sigma_1 - \sigma_z)n_1 = 0$$

Substituting the calculated principal stresses and the given values yields

$$\left(8844 - 8000\right)l_1 - 3000m_1 - 0 = 0$$
$$-3000l_1 + \left[8844 - (-4000)\right]m_1 - 3000n_1 = 0$$
$$0 - 3000m_1 + \left[8844 - (-4000)\right]n_1 = 0$$

or

$$844l_1 - 3000m_1 = 0 \Rightarrow m_1 = \frac{844}{3000}l_1 \qquad \text{(a)}$$
$$-3000l_1 + 12{,}844m_1 - 3000n_1 = 0 \qquad \text{(b)}$$
$$-3000m_1 + 12{,}844n_1 = 0 \Rightarrow n_1 = \frac{3000}{12{,}844}m_1 \qquad \text{(c)}$$

Substituting Equation (a) into Equation (c) gives

$$n_1 = \frac{3000}{12{,}844}\left(\frac{844}{3000}\right)l_1 = \frac{844}{12{,}844}l_1$$

Using Equation 4.39,

$$l_1^2 + m_1^2 + n_1^2 = 1$$

$$l_1^2 + \left(\frac{844}{3000}l_1\right)^2 + \left(\frac{844}{12{,}844}l_1\right)^2 = 1$$

or

$$l_1^2 + 0.079151l_1^2 + 0.00432l_1^2 = 1$$

Solving the above equation yields

$$1.08347l_1^2 = 1 \implies l_1 = \sqrt{\frac{1}{1.08347}} = \pm 0.961$$

Taking the positive value of l_1, we have

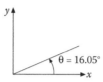

$l_1 = \cos\theta = 0.961 \implies \theta = 16.05°$ with the original x-axis.

$$m_1 = \frac{844}{3000}l_1 = \frac{844}{3000}(0.961) = 0.2704l_1$$

$m_1 = \cos\phi = 0.2704 \implies \phi = 74.31°$ with the original y-axis

$$n_1 = \frac{844}{12844}l_1 = \frac{844}{12844}(0.961) = 0.0631$$

$n_1 = \cos\psi = 0.0631 \implies \psi = 86.38°$ with the original z-axis.

4.4   STRAIN MEASUREMENT AND STRESS CALCULATIONS

To determine the state of strain ε_x, ε_y, and γ_{xy} at a point in a body, measure the normal strains as shown in Figure 4.28. Once the strains ε_a, ε_b, and ε_c are measured and their respective orientations θ_a, θ_b, and θ_c are known, the following three simultaneous equations can be used to determine ε_x, ε_y, and γ_{xy}:

$$\varepsilon_a = \varepsilon_x \cos^2\theta_a + \varepsilon_y \sin^2\theta_a + \gamma_{xy}\sin\theta_a\cos\theta_a \tag{4.40}$$

$$\varepsilon_b = \varepsilon_x \cos^2\theta_b + \varepsilon_y \sin^2\theta_b + \gamma_{xy}\sin\theta_b\cos\theta_b \tag{4.41}$$

$$\varepsilon_c = \varepsilon_x \cos^2\theta_c + \varepsilon_y \sin^2\theta_c + \gamma_{xy}\sin\theta_c\cos\theta_c \tag{4.42}$$

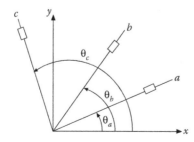

FIGURE 4.28   Normal strain measurement.

After calculating ε_x, ε_y, and γ_{xy}, the state of stresses σ_x, σ_y, and τ_{xy} and principal stresses σ_1, σ_2, and $\tau_{1,2}$ can be determined from the following equations:

$$\sigma_x = \frac{E}{\left(1 - \mu^2\right)}\left(\varepsilon_x + \mu\varepsilon_y\right) \tag{4.43}$$

$$\sigma_y = \frac{E}{\left(1 - \mu^2\right)}\left(\varepsilon_y + \mu\varepsilon_x\right) \tag{4.44}$$

$$\tau_{xy} = G\gamma_{xy} \tag{4.45}$$

1. *Principal strain equation:*

$$\varepsilon_{1,2} = \frac{\varepsilon_x + \varepsilon_y}{2} \mp \sqrt{\left(\frac{\varepsilon_x - \varepsilon_y}{2}\right)^2 + \left(\frac{\gamma_{xy}}{2}\right)^2} \tag{4.46}$$

2. *Maximum shear strain equation:*

$$\frac{\gamma_{max}}{2} = \mp\sqrt{\left(\frac{\varepsilon_x - \varepsilon_y}{2}\right)^2 + \left(\frac{\gamma_{xy}}{2}\right)^2} \tag{4.47}$$

3. *Equation for strain at any angle θ:*

$$\varepsilon_\theta = \frac{\varepsilon_x + \varepsilon_y}{2} + \left(\frac{\varepsilon_x - \varepsilon_y}{2}\right)\cos 2\theta - \frac{\gamma_{xy}}{2}\sin 2\theta \tag{4.48}$$

4. *Principal normal stress equation:*

$$\sigma_{1,2} = \frac{\sigma_x + \sigma_y}{2} \mp \sqrt{\left(\frac{\sigma_x - \sigma_y}{2}\right)^2 + \tau_{xy}^2} \tag{4.49}$$

or

$$\sigma_1 = \frac{E}{1 - \mu^2}(\varepsilon_1 + \mu\varepsilon_2) \quad \text{and} \quad \sigma_2 = \frac{E}{1 - \mu^2}(\varepsilon_2 + \mu\varepsilon_1)$$

5. *Maximum shear stress equation:*

$$\tau_{1,2} = \mp\sqrt{\left(\frac{\sigma_x - \sigma_y}{2}\right)^2 + \tau_{xy}^2} \tag{4.50}$$

6. *Equation for stress at any angle* θ:

$$\sigma_\theta = \frac{\sigma_x + \sigma_y}{2} + \left(\frac{\sigma_x - \sigma_y}{2}\right)\cos 2\theta - \tau_{xy}\sin 2\theta \qquad (4.51)$$

Example 4.8

A delta rosette at a point on the surface of loaded body, as shown in Figure 4.29, indicates the following strains:

$$\varepsilon_a = 140 \times 10^{-6} \text{ m/m}$$

$$\varepsilon_b = 212 \times 10^{-6} \text{ m/m}$$

$$\varepsilon_c = -307 \times 10^{-6} \text{ m/m}$$

Determine σ_x σ_y, τ_{xy}, and then calculate the principal normal stresses and represent them on a properly oriented element. Assume $E = 200$ GPa, $\mu = 0.25$, and $G = 80$ GPa.

SOLUTION

Using Equations 4.40 through 4.42,

$$\varepsilon_a = \varepsilon_x \cos^2 \theta_a + \varepsilon_y \sin^2 \theta_a + \gamma_{xy} \sin\theta_a \cos\theta_a$$
$$140 \times 10^{-6} = \varepsilon_x(\cos 0)^2 + 0 + 0 \implies \varepsilon_x = 140 \times 10^{-6} \qquad (a)$$

$$\varepsilon_b = \varepsilon_x \cos^2 \theta_b + \varepsilon_y \sin^2 \theta_b + \gamma_{xy} \sin\theta_b \cos\theta_b$$
$$212 \times 10^{-6} = 140 \times 10^{-6}(\cos 60)^2 + \varepsilon_y(\sin 60)^2 + \gamma_{xy} \sin 60 \cos 60 \qquad (b)$$
$$0.75\varepsilon_y + 0.433\gamma_{xy} = 177 \times 10^{-6}$$

$$\varepsilon_c = \varepsilon_x \cos^2 \theta_c + \varepsilon_y \sin^2 \theta_c - \gamma_{xy} \sin\theta_c \cos\theta_c$$
$$-307 \times 10^{-6} = 140 \times 10^{-6}(\cos 120)^2 + \varepsilon_y(\sin 120)^2$$
$$- \gamma_{xy} \sin 120 \cos 120 \qquad (c)$$
$$0.75\varepsilon_y + 0.433\gamma_{xy} = -342 \times 10^{-6}$$

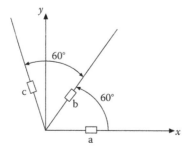

FIGURE 4.29 Delta rosette.

From Equations (b) and (c),

$$1.5\varepsilon_y = -165 \times 10^{-6} \Rightarrow \varepsilon_y = -110 \times 10^{-6}\,\text{m/m}$$

From Equation (c),

$$0.75(-110 \times 10^{-6}) + 0.433\gamma_{xy} = -342 \times 10^{-6}$$

$$\gamma_{xy} = \frac{-342 \times 10^{-6} + 82.5 \times 10^{-6}}{0.433} \Rightarrow \gamma_{xy} = 600 \times 10^{-6}$$

Using stress equations

$$\sigma_x = \frac{E}{(1 - \mu^2)}\left(\varepsilon_x + \mu\varepsilon_y\right)$$

$$\sigma_x = \frac{200 \times 10^9}{(1 - 0.25^2)}\left[140 \times 10^{-6} + (0.25)(-110 \times 10^{-6})\right]$$

$$= \frac{22500 \times 10^3}{0.9375} = 24 \times 10^6\,\text{Pa} \Rightarrow \sigma_x = 24\,\text{MPa}$$

$$\sigma_y = \frac{E}{(1 - \mu^2)}\left(\varepsilon_y + \mu\varepsilon_x\right)$$

$$\sigma_y = \frac{200 \times 10^9}{0.9375}\left[-110 \times 10^{-6} + (0.25)(140 \times 10^{-6})\right] = -16 \times 10^6\,\text{Pa}$$

$$\sigma_y = -16\,\text{MPa}$$

$$\tau_{xy} = G\gamma_{xy} = \left(80 \times 10^9\right)\left(600 \times 10^{-6}\right) = 48 \times 10^6\,\text{Pa}$$

$$\tau_{xy} = 48\,\text{MPa}$$

Principal stresses are

$$\sigma_{p1,2} = \frac{\sigma_x + \sigma_y}{2} \mp \sqrt{\left(\frac{\sigma_x - \sigma_y}{2}\right)^2 + \tau_{xy}^2}$$

$$\sigma_{p1,2} = \frac{24 - 16}{2} \mp \sqrt{\left(\frac{24 - (-16)}{2}\right)^2 + 48^2} = 4 \mp 52$$

$$\sigma_{p1} = 56\,\text{MPa}$$

$$\sigma_{p2} = -48\,\text{MPa}$$

4.5   DEFLECTION AND STIFFNESS OF BEAMS

Structural members such as beams subjected to lateral loads not only give rise to internal stresses in the beam, but also cause the beam to deflect. The deformation of a beam can be expressed in terms of the deflection of the beam from its original unloaded position. In some cases, deformation or stiffness of the structural members is more critical than the strength of the members. Deformation of CNC machine members is a good example.

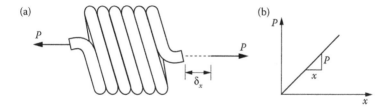

FIGURE 4.30 Spring rate and deflection. (a) Spring leaded by force P and (b) force and deflection curve.

4.5.1 SPRING RATES

The slope of the straight line between the force and the deflection curve shown in Figure 4.30b is called *spring rate*, k. The spring rate equation is

$$k = \frac{P}{x} \tag{4.52}$$

Spring rate k is constant and force in the spring is linearly proportional to its deflection. From Figure 4.30, deflection δ_x can be written as

$$\delta_x = \frac{PL}{AE} \tag{4.53}$$

Deflection can also be written as

$$\delta_x = \frac{P}{k} \tag{4.54}$$

From Equations 4.53 and 4.54,

$$k = \frac{AE}{L} \tag{4.55}$$

4.5.2 TORSION

Shafts undergo twisting called torsion. Twisting is caused by torque. The angular deformation of a uniform round bar, as shown in Figure 4.31, subjected to torque T is given as

$$\theta = \frac{TL}{GJ} \tag{4.56}$$

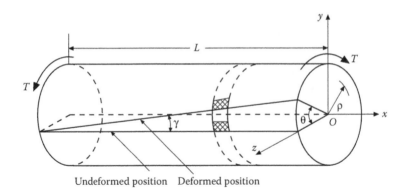

FIGURE 4.31   Angular deformation of a uniform round bar.

Note that the similarity between axial deflection and angular deformation is given by Equations 4.53 and 4.56. In Equation 4.56, J is the polar moment of inertia of the circular cross section and G is the shear modulus. Shear stress due to torque T at any distance ρ from the neutral axis of the bar shown in Figure 4.31 is given as

$$\tau = \frac{T\rho}{J} \tag{4.57}$$

The shear stress is zero at the center of the bar and maximum at the surface. Therefore, the maximum shear stress is defined as

$$\tau_{max} = \frac{Tr}{J} \tag{4.58}$$

The torsional spring rate k_t, similar to Equation 4.52, can be drawn from Equation 4.56:

$$k_t = \frac{T}{\theta} = \frac{GJ}{L} \tag{4.59}$$

4.5.2.1   Power Transmission through Rotating Shaft
In rotating systems, the power transmitted by shafts is equal to the product of the applied torque and the angular velocity of the shaft.

$$P = T\omega \tag{4.60}$$

where ω is the angular velocity in rad/s. In general, shaft speed n is given by revolutions per minute (rpm); then Equation 4.60 becomes

$$P = \frac{2\pi n T}{60} \tag{4.61}$$

Unit of power P in the US system is ft lb/s. In the SI system, unit of power is watts (N m/s or J/s). Note that one horsepower (hp) is 550 ft lb/s. In Equation 4.61, torque transmitted by the rotating shaft is given as

$$T = 6300 \frac{hp}{n} \tag{4.62}$$

where the unit of n is rpm.

Example 4.9

Determine the relative angular deformation of point O with respect to point B of the steel shaft, shown in Figure 4.32, if the applied torque is 100 N m and $G = 80$ GPa.

SOLUTION

Angular deformation of point O with respect to point B is

$$\theta_{B/O} = \theta_{A/O} + \theta_{B/A}$$

$$\theta_{B/O} = \frac{TL_{OA}}{J_{OA}G} + \frac{TL_{AB}}{J_{AB}G} = \frac{T}{G}\left(\frac{L_{OA}}{J_{OA}} + \frac{L_{AB}}{J_{AB}}\right)$$

The polar moment of inertia of solid bar members are

$$J_{OA} = \frac{\pi r^4}{2} = \frac{\pi \times (0.1/2)^4}{2} = 9.8 \times 10^{-6}\, m^4$$

$$J_{AB} = \frac{\pi r^4}{2} = \frac{\pi \times (0.05/2)^4}{2} = 0.6 \times 10^{-6}\, m^4$$

Substituting known values yields

$$\theta_{B/O} = \frac{100}{80 \times 10^9}\left(\frac{1.6}{9.8 \times 10^{-6}} + \frac{0.5}{0.6 \times 10^{-6}}\right) = 1.245 \times 10^{-3}\, rad$$

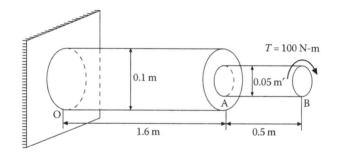

FIGURE 4.32   Steel step shaft under torsion.

Example 4.10

A solid steel shaft transmits 20 hp at a speed of 120 rpm. Determine the shaft diameter if the allowable shear stress $\tau = 5500$ lb/in.^{2} and allowable twist angle $\theta = 5°$ for each 100 in. of the shaft. Assume $G = 12 \times 10^6$ psi.

SOLUTION

Torque transmitted by the shaft is

$$T = 63{,}000 \frac{hp}{n} = 63{,}000 \frac{20}{120} = 10{,}500 \text{ lb-in.}$$

Shear stress due to torque is

$$\tau = \frac{T \times d/2}{J} \quad \text{where } J = \frac{\pi d^4}{32}$$

Substituting J into the shear stress equation yields

$$\tau = \frac{16T}{\pi d^3}$$

Then

$$5500 = \frac{16 \times 10500}{\pi d^3}$$

or

$$d^3 = \frac{16 \times 10500}{\pi \times 5500} = 9.72 \Rightarrow d = 2.14 \text{ in.}$$

Now check the design for allowable twist angle $\theta = 5°$. Rewrite the equation in degrees for the twist angle

$$\theta = \frac{TL}{JG} \times \frac{180}{\pi}$$

Rearranging yields

$$J = \frac{TL}{\theta G} \times 57.3$$

$$\frac{\pi d^4}{32} = \frac{10500 \times 100}{5 \times 12 \times 10^6} \times 57.3$$

Solving the above equation yields $d = 1.79$ in. The larger shaft diameter calculated ($d = 2.14$ in.) should be selected to satisfy the requirements of both allowable stress $\tau = 5500$ lb/in.^{2} and allowable $\theta = 5°$.

4.5.3 LATERAL DEFLECTIONS OF BEAMS

Beams are structural members subjected to transverse loads. For example, shafts carrying loads, automobile frame members, crankshafts, and building beams are designed for allowable deflection to deliver normal service. Usually, a beam requires a larger cross section to limit deflection than it does to limit stress. Many methods are available to determine beam deflections, of which the most commonly used are (1) method of integration and (2) elastic energy method.

4.5.3.1 Method of Integration

The linear differential equation relating the deflection y to the internal bending moment M of a beam is

$$EI\frac{d^2 y}{dx^2} = M \qquad (4.63)$$

where x is the axial coordinate as shown in Figure 4.33, and EI is the flexural rigidity. M represents the bending moment at the distance x from one end of the beam. The integration method to determine the beam deflection merely consists of integrating Equation 4.63. The first integration gives the slope

$$\theta = \frac{dy}{dx} \qquad (4.64)$$

and the second integration yields the deflection y corresponding to any value of x.

Example 4.11

Determine the deflection of the beam shown in Figure 4.34 by the method of integration.

SOLUTION

First draw an FBD of the structure (Figure 4.35) and express the moment equation M as a function of the coordinate x.

$$\sum F_x = 0 \Rightarrow R_{Ax} = 0$$

$$\sum F_y = 0 \Rightarrow R_{Ay} = 2.0 \text{ kN}$$

$$\sum M_A = 0 \Rightarrow M_A = 2 \times 2 = 4 \text{ kN}$$

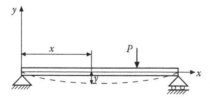

FIGURE 4.33 Beam deflection.

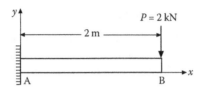

FIGURE 4.34   Cantilever beam.

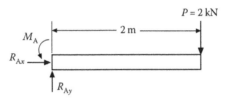

FIGURE 4.35   Free-body diagram.

To express the bending moment as a function of x, cut the beam at x distance from one end, as shown in Figure 4.36, and apply the equilibrium equations. Equilibrium at any distance x is

$$M_x + 4 \times 10^3 - 2 \times 10^3 x = 0$$

or

$$M_x = 2 \times 10^3 x - 4 \times 10^3$$

Recalling

$$M_x = EI\frac{d^2y}{dx^2}$$

then

$$EI\frac{d^2y}{dx^2} = 2 \times 10^3 x - 4 \times 10^3$$

Integrating the above equation gives

$$EI\frac{dy}{dx} = 2 \times 10^3 \frac{x^2}{2} - 4 \times 10^3 x + C_1 \tag{a}$$

Integrating a second time yields

$$EIy = 2 \times 10^3 \frac{x^3}{2 \times 3} - 4 \times 10^3 \frac{x^2}{2} + C_1 x + C_2 \tag{b}$$

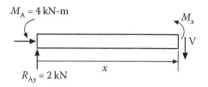

FIGURE 4.36 Section of beam at any distance x.

Using the following boundary conditions, find C_1 and C_2

$$\text{Slope} \Rightarrow \frac{dy}{dx}\Big|_{x=0} = 0 \quad \text{and} \quad \text{deflection} \Rightarrow y\big|_{x=0} = 0$$

From Equation (a),

$$EI\left(\frac{dy}{dx} = 0\right) = 2 \times 10^3(0) - 4 \times 10^3(0) + C_1 = 0 \Rightarrow C_1 = 0$$

From Equation (b),

$$EI\left(y = 0\right) = 2 \times 10^3(0) - 4 \times 10^3(0) + C_1(0) + C_2 = 0 \Rightarrow C_2 = 0$$

Substituting C_1 and C_2 into Equation (b) yields

$$y = \frac{10^3}{EI}\left(-2x^2 + \frac{1}{3}x^3\right)$$

4.5.3.2 Elastic Energy Method (Castigliano's Method)

Consider an elastic bar with a tip weight, W, at the end of the bar as shown in Figure 4.37a. Due to the weight, W, the bar will deform δ amount. Within the elastic region, $W = F$ is proportional to δ which can be expressed as

$$F = k\delta \tag{4.65}$$

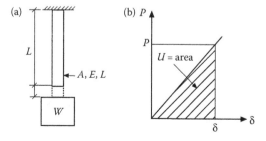

FIGURE 4.37 Elastic bar with a tip weight. (a) Elastic bar loaded by weight, w and (b) force and deflection curve.

As shown in Figure 4.37b, the work done by F is $1/2P\delta$. Thus, the strain energy stored in the bar is the area of the triangle shown in Figure 4.37b.

$$U = \frac{1}{2}P\delta \tag{4.66}$$

But $\delta = PL/AE$ and substituting in the strain energy equation yields

$$U = \frac{1}{2}\frac{P^2L}{AE} \tag{4.67}$$

The strain energy expression in tension or compression of a bar with length L, with a varying cross section and possibly a varying modulus of elasticity, is

$$U = \mp\int_0^L \frac{P^2}{2EA}\,dx \tag{4.68}$$

Strain energy due to the bending moment is

$$U = \int_0^L \frac{M^2\,dx}{2EI} \tag{4.69}$$

Strain energy due to torsion is

$$U = \int_0^L \frac{T^2\,dx}{2GJ} \tag{4.70}$$

Strain energy due to transverse shear (for a rectangular section) is

$$U = \int_0^L \frac{3V^2\,dx}{5GA} \tag{4.71}$$

In the above strain energy equations, I is the moment of inertia, E is the modulus of elasticity, J is the polar moment of inertia, and G is the modulus of rigidity.

As discussed before, a structure subjected to external forces absorbs energy in the form of strain energy as expressed in Equations 4.68 through 4.71. Castigliano's method uses strain energy to determine deflections of the structure. This method states that the partial derivative of the strain energy of a structure with respect to concentrated applied load is equal to the deflection of the structure at the point of application of the load. The direction of the deflection is in the direction of the applied load. If the applied load is a force, then the deflection is translational. The rotational deflection can be calculated if the applied load is a moment or torque. Hence, the equation for translational deflection is

$$y = \frac{\partial U}{\partial Q} \tag{4.72}$$

where Q is the concentrated applied load. Rotational deflection is given by

$$Q = \frac{\partial U}{\partial M}$$

(4.73)

where M is the applied moment or torque.

Example 4.12

A constant cross-sectional steel beam, as shown in Figure 4.38a, is subjected to applied load $P = 4000$ lb. Using the elastic energy method, determine the lateral deflection of the beam at point B. Assume that the beam has a circular cross section and $E = 30 \times 10^6$.

SOLUTION

First calculate the reaction forces from the FBD shown in Figure 4.38b.

$$\sum F_x = 0 \Rightarrow R_{Ax} = 0$$

$$\sum F_y = 0 \Rightarrow R_{Ay} + R_{Cy} = P$$

$$\sum M_A = 0$$

$$100 \times R_{Cy} - 60 \times P = 0 \Rightarrow R_{Cy} = \frac{3}{5}P$$

and

$$R_{Ay} = \frac{2}{5}P$$

(a)

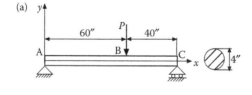

(b)

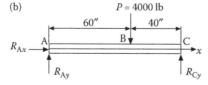

FIGURE 4.38 (a) Beam under applied load and (b) free-body diagram.

To define moment M_x as a function of the coordinate x and P within the region A–B, cut the beam at x distance between A and B and apply the equilibrium equations (Figure 4.39a).

$$\sum F_y = 0$$

$$2/5\,P - V = 0 \Rightarrow V = \frac{2}{5}\,P$$

$$\sum M_x = 0 \Rightarrow M_x = \frac{2}{5}P \times x$$

Similarly, to find the bending moment as a function of x and P within region B–C, cut the beam at any point between B and C; and apply the equilibrium equations (Figure 4.39b).

$$\sum F_y = 0$$

$$\frac{2}{4}P - P - V = 0 \Rightarrow V = -\frac{3}{5}P$$

$$\sum M_x = 0$$

$$M_x - \frac{2}{5}Px + P(x - 60) = 0$$

or

$$M_x = -\frac{3}{5}Px + 60P$$

Since the beam is long, shear stress can be ignored. Thus, consider the bending stress for the strain energy equation

$$U = \frac{1}{2EI}\int_0^L M^2\,dx$$

$$U = \frac{1}{2EI}\left\{\int_0^{60}\left(\frac{2}{5}Px\right)^2 dx + \int_{60}^{100}\left(-\frac{3}{5}Px + 60P\right)^2 dx\right\}$$

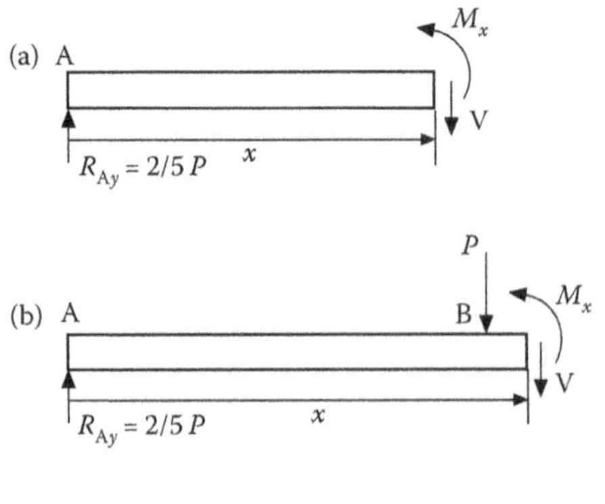

(a)

(b)

FIGURE 4.39 (a) Free-body diagram between distance A and B and (b) free-body diagram between distance points B and C.

Deflection at B is

$$Y_B = \frac{\partial U}{\partial P}$$

Applying the above equation for the first part of the strain energy equation yields

$$\frac{\partial U}{\partial P}\bigg|_{part1} = \int_0^{60}\left(\frac{8}{25}Px^2\right)dx$$

Integrating yields

$$\frac{\partial U}{\partial P}\bigg|_{part1} = \frac{8}{25}P\frac{x^3}{3}\bigg|_0^{60} = 23{,}040\,P$$

For the second part of the strain energy equation

$$\frac{\partial U}{\partial P}\bigg|_{part2} = \int_{60}^{100}\left(+\frac{9}{25}2Px^2 - 144Px + 7200P\right)dx$$

Integrating yields

$$\left(\frac{9}{25}2P\frac{x^3}{3} - 144P\frac{x^2}{2} + 7200Px\right)\bigg|_{60}^{100} = 15{,}360P$$

Therefore,

$$\frac{\partial U}{\partial p}\bigg|_{all} = \frac{1}{2EI}(23{,}040P + 15{,}360P) = \frac{19{,}200P}{EI}$$

Substituting

$$E = 30 \times 10^6$$

$$I = \frac{\pi d^4}{64} = \frac{\pi 4^4}{64} = 12.57\,\text{in.}^4$$

yields

$$Y_B = \frac{\partial u}{\partial p}\bigg|_B = \frac{19{,}200 \times 4000}{30 \times 10^6 \times 12.57} = 0.204\,\text{in.}$$

Example 4.13

Determine the beam deflection in Figure 4.40a at point D, 25″ to the right from point A.

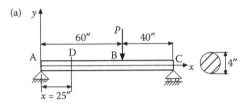

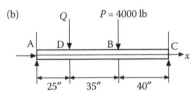

FIGURE 4.40   (a) Beam loaded by load *p* and (b) beam free-body diagram.

SOLUTION

Since there is no concentrated load at the point where $x = 25''$, introduce a dummy load Q at the point where the deflection is desired. After the strain energy equation is obtained, differentiate with respect to the dummy load Q. Finally, set the dummy load Q equal to zero to determine the deflection value.

First draw an FBD (Figure 4.40b) and determine the reaction forces.

$$\sum F_y = 0 \Rightarrow R_{Ay} - Q - P + R_{Cy} = 0$$

or

$$R_{Ay} + R_{Cy} = Q + P$$

$$\sum M_A = 0 \Rightarrow -25Q - 60P + 100R_{Cy} = 0 \tag{a}$$

or

$$R_{Cy} = \frac{3}{5}P + \frac{1}{4}Q \tag{b}$$

Substituting Equation (b) into Equation (a) gives

$$R_{Ay} = P + Q - \frac{1}{4}Q - \frac{3}{5}P = \frac{5P - 3P}{5} + \frac{4Q - Q}{4}$$

or

$$R_{Ay} = \frac{2}{5}P + \frac{3}{4}Q \tag{c}$$

Find the bending moment as a function of x, P, and Q within the region of A–D (Figure 4.41a).

$$\sum M_x = 0 \Rightarrow M_x - \left(\frac{2}{5}P + \frac{3}{4}Q\right)x_1 = 0$$

or

$$M_x = \left(\frac{2}{5}P + \frac{3}{4}Q\right)x_1 \quad \text{for} \quad 0 < x_1 < 25''$$

(d)

Find the bending moment as a function of x, P, and Q within the region of D–B (Figure 4.41b).

$$\sum M_x = 0 \Rightarrow M_x = \left(\frac{2}{5}P + \frac{3}{4}Q\right)x_1 - Q(x_1 - 25'') \quad \text{for} \quad 25'' < x_1 < 60''$$

Find the bending moment as a function of x, P, and Q within the region of B–C (Figure 4.41c).

$$\sum M_x = 0 \Rightarrow M_x = \left(\frac{3}{5}P + \frac{1}{4}Q\right)x_2 \quad \text{for} \quad 0 < x_2 < 40''$$

Then, strain energy is

$$U = \frac{1}{2EI}\left\{\int_0^{25}\left(\frac{2}{5}P + \frac{3}{4}Q\right)^2 x_1^2\,dx_1 + \int_{25}^{60}\left[\left(\frac{2}{5}P + \frac{3}{4}Q\right)x_1 - Q(x_1 - 25)\right]^2 dx_1\right.$$

$$\left. + \int_0^{40}\left(\frac{3}{5}P + \frac{1}{4}Q\right)^2 x_2^2\,dx_2\right\}$$

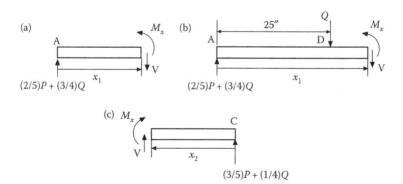

FIGURE 4.41 (a) Free-body diagram between points A and D, (b) free-body diagram between points A and B, and (c) free-body diagram between points C and B.

To determine the deflection at $x = 25''$ where the dummy load Q is applied, perform the following steps:

Step 1: Take the partial derivatives with respect to Q.

$$Y_{D=25''} = \frac{\partial U}{\partial Q} = \frac{1}{EI}\int_0^{25}\left(\frac{2}{5}P + \frac{3}{4}Q\right)\frac{3}{4}x_1^2\,dx_1$$

$$+ \frac{1}{EI}\int_{25}^{60}\left[\left(\frac{2}{5}P + \frac{3}{4}Q\right)x_1 - Q(x_1 - 25)\right]\left[\frac{3}{4}x_1 - (x_1 - 25)\right]dx_1$$

$$+ \frac{1}{EI}\int_0^{40}\left(\frac{3}{5}P + \frac{1}{4}Q\right)\frac{1}{4}x_2^2$$

Step 2: Set $Q = 0$.

$$\frac{\partial U}{\partial Q}\bigg|_{Q=0} = \frac{1}{EI}\int_0^{25}\frac{3}{10}Px_1^2\,dx_1 + \frac{1}{EI}\int_{25}^{60}\frac{2}{5}Px_1\left[\frac{3}{4}x_1 - (x_1 - 25)\right]dx_1$$

$$+ \frac{1}{EI}\int_0^{40}\frac{3}{20}Px_2^2\,dx_2$$

Step 3a: Integrate Part I of the resulting equation.

$$\frac{1}{EI}\int_0^{25}\frac{3}{10}Px_1^2\,dx_1 = \frac{1}{EI}\left|\frac{3}{10}P\frac{x_1^3}{3}\right|_0^{25}$$

$$= \frac{1}{30\times10^6\times12.57}\left|\frac{3}{10}\times4000\times\frac{25^3}{3}\right| = 0.0166''$$

Step 3b: Integrate Part II of the resulting equation.

$$\frac{1}{EI}\int_{25}^{60}\frac{2}{5}Px_1\left[\frac{3}{4}x_1 - (x_1 - 25)\right]dx_1$$

$$= \frac{1}{EI}\int_{25}^{60}\left(\frac{3}{10}Px_1^2 - \frac{2}{5}Px_1^2 + 10Px_1\right)dx_1$$

$$= \frac{1}{EI}\left|\frac{3}{10}P\frac{x_1^3}{3} - \frac{2}{5}P\frac{x_1^3}{3} + 10P\frac{x_1^2}{2}\right|_{25}^{60}$$

$$= \frac{1}{EI}\left[\left(\frac{1}{10}P60^3 - \frac{2}{15}60^3P + 5P60^2\right) - \left(\frac{1}{10}P25^3 - \frac{2}{15}P25^3 + 5P25^2\right)\right]$$

$$= \frac{1}{30\times10^6\times12.57}\left[(21,600\times4000 - 28,800\times4000 + 18,000\times4000)\right.$$

$$\left. - (1562.5\times4000 - 2083.3\times4000 + 3125\times4000)\right]$$

$$= 0.0869''$$

Step 3c: Integrate Part III of the resulting equation.

$$\frac{1}{EI}\int_0^{40}\frac{3}{20}Px_2^2\,dx_2 = \frac{1}{EI}\left.\frac{3}{20}P\frac{x_2^3}{3}\right|_0^{40} = \frac{1}{EI}\left(\frac{3}{20}P\frac{40^3}{3}\right) = \frac{1}{EI}(3200P)$$

$$= \frac{3200 \times 4000}{30 \times 10^6 \times 12.57} = 0.0339''$$

Thus total deflection at x is

$$Y_{D=25''} = 0.0166'' + 0.0869'' + 0.0339'' = 0.1374''$$

4.5.3.3 Statically Indeterminate Structures: Solution of Reactions and Deflections by Castigliano's Method

Castigliano's method can be used to determine the reactions in a statically indeterminate structure. A structure is called statically indeterminate if the number of unknown reactions is more than the number of independent equations of equilibrium. It is thus necessary to obtain additional equations to solve the unknowns. Additional independent equations can be obtained from

$$\frac{\partial u}{\partial R_j} = \frac{\partial u}{\partial F_j} \tag{4.74}$$

where R_j denotes the structure reaction forces and F_j denotes the structure member forces. Castigliano's method can only be used for the condition in which displacement is proportional to the force that produced it.

Example 4.14

A rigid horizontal bar AD is pinned at A and supported by a steel wire at B and steel bar at C, as shown in Figure 4.42. Determine the reactions at A, the forces in steel bar and steel wire, and the deflection due to the 10,000-lb load at D.

SOLUTION

Apply equilibrium equations

$$\sum F_x = 0 \Rightarrow R_{Ax} = 0 \tag{a}$$

$$\sum F_y = 0 \Rightarrow R_{Ay} + T + F_c - 10,000 = 0 \tag{b}$$

$$\sum M_A = 0 \Rightarrow 40T + 100F_c = 150P \tag{c}$$

In the above equations there are three unknowns which require another equation to solve the unknowns. Since forces at support A, in the wire, and in the two-force

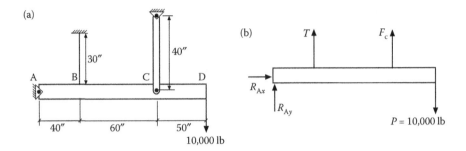

FIGURE 4.42   (a) Statically indeterminate structure and (b) free-body diagram.

member CE do not move, there are no changes in strain energy with respect to these forces. Therefore,

$$\frac{\partial u}{\partial R_{Ay}} = 0; \quad \frac{\partial u}{\partial T} = 0; \quad \frac{\partial u}{\partial F_c} = 0 \tag{d}$$

The strain energy stored by the wire and the two-force member CE is

$$U = U_{\text{wire}} + U_{\text{member CE}}$$

$$U = \left(\frac{T^2 L}{2AE}\right)_{\text{wire}} + \left(\frac{F_c^2 L}{2AE}\right)_{\text{member CE}}$$

$$U = \left(\frac{T^2 \times 30}{2 \times 30 \times 10^6 \times 1}\right) + \left(\frac{F_c^2 \times 40}{2 \times 2 \times 30 \times 10^6}\right)$$

or

$$U = \left(\frac{T^2}{20 \times 10^5}\right) + \left(\frac{F_c^2}{30 \times 10^5}\right)$$

Note that since bar A–D is rigid, it does not store strain energy.
    Using Equation (d) yields

$$\frac{\partial U}{\partial T} = 0 \implies \left(\frac{2T}{20 \times 10^5}\right) + \left(\frac{2F_c}{30 \times 10^5}\right)\left(\frac{\partial F_c}{\partial T}\right) \tag{e}$$

Note that F_c is a function of T (see Equation (c). From Equation (c)

$$10F_c = 15P - 4T \implies F_c = \frac{15}{10}P - \frac{4}{10}T \tag{f}$$

$$\frac{\partial F_c}{\partial T} = -\frac{4}{10} = -\frac{2}{5}$$

Then

$$\frac{\partial U}{\partial T} = \frac{T}{10 \times 10^5} + \left(\frac{F_c}{15 \times 10^5}\right)\left(-\frac{2}{5}\right) = 0$$

(g)

$$T = 10 \times 10^5 \frac{F_c}{15 \times 10^5} \times \frac{2}{5} = 0.266F_c$$

Solving for R_{Ay}, T, and F_c from Equations (b), (c), and (g) gives

$$4(0.266F_c) + 100F_c = 150(10000) \;\Rightarrow\; F_c = 13557.5\,\text{lb}$$

and

$$T = 0.266 \times 13557.5 = 3606.3\,\text{lb}$$

then

$$R_{Ay} = 10{,}000 - 3606.3 - 13557.5 = -7163.8\,\text{lb}$$

Use Castigliano's method to determine the deflection at point D:

$$U = \frac{T^2}{20 \times 10^5} + \frac{F_c^2}{30 \times 10^5}$$

$$Y_D = \frac{\partial U}{\partial P} = \frac{\partial U}{\partial T} \times \frac{\partial T}{\partial P} + \frac{\partial U}{\partial F_c} \times \frac{\partial F_c}{\partial P}$$

or

$$Y_D = \frac{T}{10 \times 10^5} \times \frac{\partial T}{\partial P} + \frac{F_c}{15 \times 10^5} \times \frac{\partial F_c}{\partial P}$$

Forces T and F_c can be written in terms of P from Equations (c) and (g) as

$$4T + 10\left(\frac{T}{0.266}\right) = 15P$$

$$41.6T = 15P \;\Rightarrow\; T = \frac{15}{41.6}P$$

Substituting yields

$$10F_c = 15P - 4\left(\frac{15}{41.6}P\right) \;\Rightarrow\; F_c = 1.356P$$

Hence,

$$\frac{\partial T}{\partial P} = \frac{15}{41.6}$$

and

$$\frac{\partial F_c}{\partial P} = 1.356$$

Substituting into the deflection equation

$$Y_D = \frac{T}{10 \times 10^5} \times \frac{15}{41.6} + \frac{F_c}{15 \times 10^5} \times 1.356$$

$$Y_D = \frac{3606.3}{10 \times 10^6} \times \frac{15}{41.6} + \frac{13,557.5}{15 \times 10^5} \times 1.356 = 0.0135''$$

4.6 BUCKLING OF COLUMNS

A long compression member that supports buildings, bridges, or similar structures is called a column. Columns totally collapse or buckle under the action of bending stress or a combination of bending and direct compression. The compression load at the onset of buckling is called the critical load, P_{cr}. Columns or compression members can be divided into three groups: (1) short columns or posts, (2) intermediate columns, and (3) long slender columns.

Slenderness ratio L/k is used as the criterion to classify the three groups of columns. L is the unsupported column length and k is the radius of gyration, where

$$k = \sqrt{\frac{I}{A}} \tag{4.75}$$

In Equation 4.75, I is the least moment of inertia and A is the cross-sectional area of the column.

For an ideal long slender column, as shown in Figure 4.43, the following Euler equation is used to determine the critical load:

$$P_{cr} = \frac{\pi^2 EI}{L^2} \tag{4.76}$$

The limitation of validity of the Euler critical load to a slender column is given as

$$\frac{L}{k} \geq \sqrt{\frac{2\pi^2 E}{\sigma_y}} \tag{4.77}$$

When slenderness ratio becomes

$$\frac{L}{k} < \sqrt{\frac{2\pi^2 E}{\sigma_y}} \tag{4.78}$$

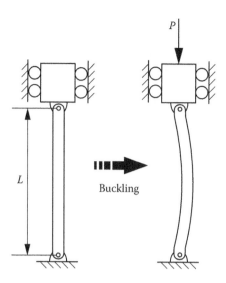

FIGURE 4.43   Ideal pinned–pinned column.

then assumptions for short columns should be used and Johnson's formula should be applied to determine the critical load. Johnson's formula is given as

$$\frac{P_{cr}}{A} = \sigma_y - \frac{1}{E}\left(\frac{\sigma_y L}{2\pi k}\right)^2 \tag{4.79}$$

The ideal column equation is only valid for a column with both ends pinned. Since the end conditions will affect the critical buckling load, the ideal equation given by Equation 4.76 must be modified if the ends are not pinned.

For columns with different types of supports (end conditions), Euler's formula may still be used if the distance L is replaced with the effective length, L_{ef}. Hence, the critical load equation becomes

$$P_{cr} = \frac{\pi^2 EI}{L_{ef}^2} \tag{4.80}$$

where L_{ef} is the effective length of a column which takes into account column length and the end conditions. Figure 4.44 shows the L_{ef} values for different end conditions.

Example 4.15

A steel solid round bar with a diameter of 0.05 m pinned at one end and fixed at the other end is subjected to compressive load. Calculate the critical load for buckling. Assume that $L = 2$ m, $E = 200$ GPa, and $\sigma_y = 689$ MPa.

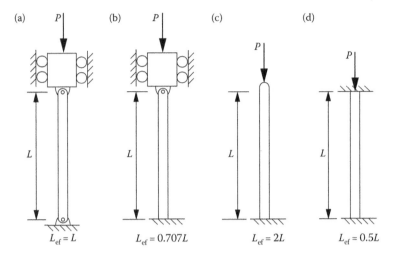

FIGURE 4.44 Column end conditions and effective length values. (a) Roller and pinned end, (b) roller and fixed end, (c) fixed and free end, and (d) both fixed end conditions.

SOLUTION

$$I = \frac{\pi r^4}{4} = \frac{\pi(0.05/2)^4}{4} = 3.06 \times 10^{-7}\,\text{m}^4$$

$$A = \frac{\pi D^2}{4} = \frac{\pi \times 0.05^2}{4} = 1.96 \times 10^{-3}\,\text{m}^2$$

Substituting known values in

$$k = \sqrt{\frac{I}{A}}$$

gives

$$k = \sqrt{\frac{3.06 \times 10^{-7}}{1.96 \times 10^{-3}}} = 0.012$$

Then

$$\frac{L}{k} = \frac{2}{0.012} = 166.6$$

Check the limitation of validity of the Euler critical load for the slender column by applying

$$\sqrt{\frac{2\pi^2 E}{\sigma_y}} = \sqrt{\frac{2\pi^2\,200 \times 10^9}{689 \times 10^6}} = 75.65$$

Since

$$\frac{L}{k} = 166.6 \geq 75.65$$

the Euler equation is applicable. Then the critical load is

$$P_{cr} = \frac{\pi^2 EI}{L_{ef}^2} = \frac{\pi^2 (200 \times 10^9) 3.06 \times 10^{-7}}{(0.707 \times 2)^2} = 3018 \times 10^2 \, \text{N}$$

Note that in the above equation $L_{ef} = 0.707 \, L$.

4.6.1 ECCENTRICALLY LOADED COLUMN

If the load P shown in Figure 4.44 has an eccentricity e, the Secant formula can be used to determine the maximum axial load P that may be eccentrically applied to a given column,

$$\sigma_y = \frac{NP}{A} \left(1 + \frac{ec}{k^2} \sec \frac{L_e}{2} \sqrt{\frac{NP}{IE}} \right)$$

where c is the distance from the centroidal axis to the external fiber, σ_y is the yield strength of the material, and N is the safety factor.

4.6.2 PRESTRESSED CONCRETE COLUMN

Concrete shows good strength when it is subjected to compressive loads. However, concrete strength in tension is very low, almost one-tenth of its compressive strength. Prestressing is the creation of permanent stresses in concrete columns to prevent cracking and its strength behavior under various service conditions.

Prestressed concrete columns are a design method to provide tensile strength to concrete columns by using tendons (wires, strands, or steel bars) in small ducts. A threaded tendon is anchored to one end of the concrete and tension is applied from the other end so that the tensile force is transferred to the concrete as compressive stress (see Figure 4.45). Hence, when the concrete column is subjected to tensile loads the initial compressive state of the concrete column will be released and it will not experience tensile stress. This will prevent the concrete column from cracking under tensile loads.

Prestressed concrete is one of the most reliable, durable, and widely used construction materials in the construction industry. It is used in buildings, bridges, nuclear power vessels, TV towers, offshore drilling platforms, and so on.

Figure 4.46 shows the FBD of a concrete column having a single prestress tendon at the center of the column. As shown in the figure, the concrete column is under compression due to axial force F. The bending stress σ_b^{con} due to the bending moment M is

$$\sigma_b^{con} = \pm \frac{Mc}{I} \tag{4.81}$$

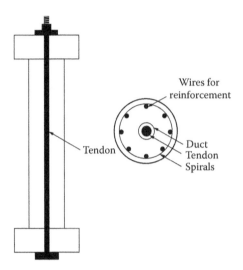

FIGURE 4.45 Prestressed concrete column.

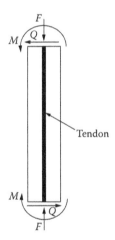

FIGURE 4.46 Free-body diagram.

where c is the distance from the centroidal axis to the outer surface of the column and I is the moment of inertia. The axial stress σ_a^{con} due to the compressive load is

$$\sigma_a^{con} = -\frac{F}{A} \tag{4.82}$$

where A is the total cross-sectional area. Compressive stress σ_p^{ten} due to tensile force in the tendon (prestress) is

$$\sigma_p^{ten} = -\frac{T}{A} = -\frac{\sigma_s^{ten} \times A^{ten}}{A} \tag{4.83}$$

where σ_s^{ten} is the tensile prestress in the tendon and A^{ten} is the cross-sectional area of the tendon. Then the resulting stress σ_R^{con} in the concrete column is

$$\sigma_R^{con} = \pm \frac{Mc}{I} - \frac{F}{A} - \frac{\sigma_s^{ten} \times A^{ten}}{A} \tag{4.84}$$

Equation 4.84 can be modified to determine the cross-sectional area of the tendon A^{ten} required to eliminate the effect of tensile stress due to bending. For this, Equation 4.84 should be equal to zero. Then Equation 4.84 becomes

$$A^{ten} = \frac{A}{\sigma_s^{ten}} \left(\frac{Mc}{I} - \frac{F}{A} \right) \tag{4.85}$$

Once the cross-sectional area of the tendon is known, the maximum compressive stress in the concrete column can be determined from

$$\sigma_{max}^{con} = -\frac{Mc}{I} - \frac{F}{A} - \frac{\sigma_s^{ten} \times A^{ten}}{A} \tag{4.86}$$

Compressive stress given by Equation 4.86 cannot be larger than 45% of the compressive strength of the concrete, S_{comp}^{con}. Then the allowable compressive stress σ_{all}^{con} is

$$\sigma_{all}^{con} = 0.45 \times S_{comp}^{con}$$

The principal tensile stress σ_1 can be calculated from

$$\sigma_1 = \frac{\sigma}{2} + \sqrt{\left(\frac{\sigma}{2} \right)^2 + \tau^2} \tag{4.87}$$

where τ is the shear stress at any point and σ is the corresponding normal stress. In general, the critical fracture stress $\sigma_{cr\text{-}frac}^{con}$ is assumed to be

$$\sigma_{cr\text{-}frac}^{con} = 4\sqrt{S_{comp}^{con}} \tag{4.88}$$

where S_{comp}^{con} is the compressive strength of the concrete column. Then, the condition to prevent shear cracking can be defined as

$$\sigma_1 \leq 4\sqrt{S_{comp}^{con}} \tag{4.89}$$

When the concrete column is long, the uniaxial critical cracking stress based on the modulus of rapture can be estimated as

$$\sigma_{cr\text{-}crack}^{con} = 7.5\sqrt{S_{comp}^{con}} \tag{4.90}$$

More information on this subject is given by Dawson [1].

PROBLEMS

4.1. Determine the total deformation of the straight steel bar shown in the following figure. Assume the modulus of elasticity $E = 200$ GPa, $L = 0.3$ m, and $P = 35.6$ kN.

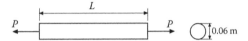

4.2. Determine the total deformation of the aluminum hollow tube shown in the following figure. Assume the modulus of elasticity $E = 200$ GPa and $L = 0.4$ m.

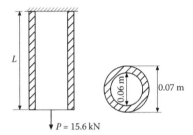

4.3. In Example 4.1, if it is assumed that wall B is not rigid and is moved 0.3×10^{-3} m to the left, determine the stress in the rod.

4.4. The steel bar shown in the following figure is in tension due to an applied force of 150 kN. If the measured elongation of the rod is 0.3 mm and the decrease in diameter is 0.015 mm, determine Poisson's ratio of the material.

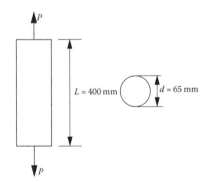

4.5. Two solid bars with circular cross-sectional areas are rigidly fastened at b–b and welded at a–a to support an applied load of 12,000 lb as shown in the following figure. Determine the maximum stress in each material.

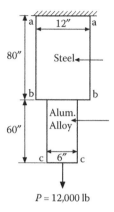

$P = 12,000$ lb

4.6. The square cross-sectional member AB $(0.9 \times 10^{-3}$ m²) shown in the following figure is made of steel. Assume the modulus of elasticity $E = 200$ GPa. Determine the allowable force, F, applied to meet the following requirements:
a. The maximum tensile stress in AB must not exceed 345 MPa.
b. The elastic deformation of member AB must not exceed 0.5×10^{-3} m.

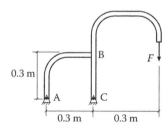

4.7. As shown in following figure, a bar of length L is supported by a steel bar and an aluminum bar. If the allowable stress levels in the steel and aluminum bars are $\sigma_{st} = 40$ kpsi and $\sigma_{al} = 24$ kpsi, respectively, determine the stresses in the steel and aluminum bars. Ignore the weight of the bar which carries 10,000 lb and assume this bar is rigid. The cross-sectional areas of the steel and aluminum bars are 1 in.² and 2 in.², respectively. Assume $E = 30 \times 10^6$ psi for steel and $E = 10 \times 10^6$ psi for aluminum.

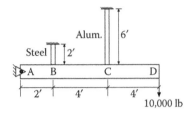

4.8.   Assume that a brass bar is inside the aluminum cylinder as shown
       in the following figure. The brass bar is 0.002 in. longer than the alu-
       minum cylinder. If the allowable stress levels for brass and aluminum
       are $\sigma_{br} = 20$ kpsi and $\sigma_{al} = 24$ kpsi, respectively, determine the maxi-
       mum force, P, that can be applied by the hydraulic cylinder. Assume
       $E_{al} = 10 \times 10^6$ psi for aluminum and $E_{br} = 12 \times 10^6$ psi for brass. The
       cross-sectional areas of the aluminum and brass bars are 6 in.^{2} and 2 in.^{2},
       respectively.

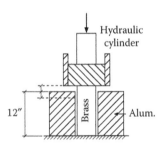

4.9.   A weight, W, is supported by two brass bars and one aluminum bar
       as shown in the following figure. If the allowable stresses in the brass
       and aluminum bars are $\sigma_{br} = 24$ kpsi and $\sigma_{al} = 20$ kpsi, respectively,
       determine the maximum weight that can be supported by the three
       bars. Assume $E_{al} = 10 \times 10^6$ psi for aluminum and $E_{br} = 12 \times 10^6$ psi
       for brass. The cross-sectional areas of the aluminum and brass bars are
       2 in.^{2} and 1 in.^{2}, respectively.

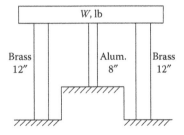

4.10. Referring to Example 4.6, draw an element to show the corresponding normal stresses with respect to the y-axis on the original element.

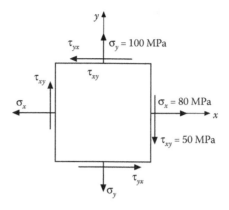

4.11. Two solid bars made of steel and aluminum are rigidly fixed to walls A and C as shown in the following figure. At point B where the two bars are attached, $T = 8,000$ lb in. is applied. Determine the maximum shear stresses in the steel and aluminum bars. Assume $G_{st} = 12 \times 10^6$ psi for steel and $G_{al} = 3.9 \times 10^6$ psi for aluminum.

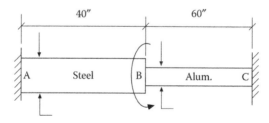

4.12. As shown in the following figure, the strains on the outside surface of a thin-walled cylindrical vessel are

$$\varepsilon_a = 4 \times 10^{-4} \text{ m/m}$$
$$\varepsilon_b = 4 \times 10^{-4} \text{ m/m}$$
$$\varepsilon_c = 6 \times 10^{-4} \text{ m/m}$$

Assuming $E = 200$ GPa, $\mu = 0.25$, and $t = 0.005$ m (wall thickness and inside radius), determine the principal normal stresses.

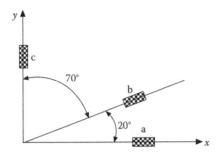

4.13. Determine the maximum allowable load, W, of the beam shown in the following figure if the maximum allowable deflection of the center of the steel beam is 0.4 in. Assume $E = 30 \times 10^6$ psi and $I = 100$ in.^{4}.

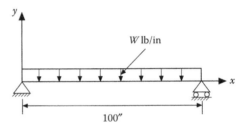

4.14. The shaft and gear system shown in the following figure is driven by a motor that supplies 5 kW power at a constant speed of 1500 rpm. Determine the maximum shear stress in shaft AB.

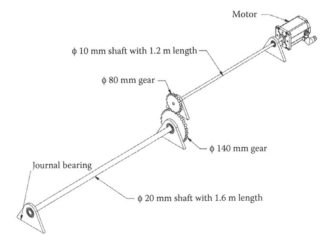

4.15. A 0.8-m-long piston rod used in steam engine is subjected to an axial load of 75,000 N. Determine the safe diameter if a safety factor of 2 is assumed. Use $\sigma_y = 689$ MPa and $E = 200$ GPa for material properties of the piston rod. Assume $L_e = L$ (pinned–pinned connection).

REFERENCE

1. Dawson, T. H. 1983. *Offshore Structural Engineering*, Prentice-Hall, Englewood Cliffs, NJ.

BIBLIOGRAPHY

Bauld, N. R. *Mechanics of Materials*, Brooks/Cole, Monterey, CA, 1982.

Higdon, A., Ohlsen, H. E., Stiles, B. W., Weese J. A., and Riley, F. W. *Mechanics of Materials*, John Wiley & Sons. New York, 1985.

Fletcher, Q. D. *Mechanics of Materials*, Holt, Rinehart and Winston, New York, 1985.

Popov, P. E., and Balan, A. T. *Engineering Mechanics of Solids*, Prentice-Hall, Englewood Cliff, NJ, 1999.

5 Failure Theories and Dynamic Loadings

5.1 FAILURE THEORIES AND SAFETY FACTOR

In designing machine components for safe design, design engineers make sure that the internal stresses do not exceed the strength of the material. Ductile materials fail by yielding (where S_y is the criterion of failure) and brittle materials fail when there is a local fracture or rupture (where S_{ut} is the criterion of failure).

If structural members are subject only to simple stresses such as tension or compression, the determination of safe-design stress is a simple task. However, most structural members are subjected to a combination of stresses as discussed earlier. It is not safe to design structural members on the basis of simple-stress experiments when they are subjected to combined stresses. Safety factors based on several failure theories are discussed below.

5.1.1 Maximum Normal Stress Theory

According to the maximum normal stress theory, failure occurs whenever the largest principal stress equals the yield strength.

$$\sigma = S_y \tag{5.1}$$

The safety factor is defined as

$$n = \frac{S_y}{\sigma} \tag{5.2}$$

5.1.2 Maximum Shear Stress Theory

This theory states that failure occurs whenever the maximum resultant shear stress, τ_{max}, equals the shear strength at the yield point of the material in a simple tensile test.

$$\tau_{max} = \tau_y \tag{5.3}$$

where τ_y is the shear strength at the instant of failure by yielding during the tensile test. Thus,

$$\tau_y = \frac{1}{2} S_y \tag{5.4}$$

Using this theorem, the safety factor is defined as

$$n = \frac{\tau_y}{\tau_{max}} = \frac{S_y}{2\tau_{max}} \tag{5.5}$$

5.1.3 DISTORTION ENERGY THEORY

The distortion energy theory, also known as the Von Mises theory, assumes that yielding failure occurs when the distortion energy equals the yield strength of the material in a simple tensile test. This theory is mathematically shown as

$$S_y^2 = \sigma_1^2 + \sigma_2^2 + \sigma_3^2 - \sigma_1\sigma_2 - \sigma_2\sigma_3 - \sigma_1\sigma_3 \tag{5.6}$$

The safety factor is defined as

$$n = \frac{S_y}{\sqrt{\sigma_1^2 + \sigma_2^2 + \sigma_3^2 - \sigma_1\sigma_2 - \sigma_2\sigma_3 - \sigma_1\sigma_3}} \tag{5.7}$$

For a two-dimensional stress field, $\sigma_3 = 0$. Thus, Equation 5.7 becomes

$$n = \frac{S_y}{\sqrt{\sigma_1^2 + \sigma_2^2 - \sigma_1\sigma_2}} \tag{5.8}$$

The allowable stress, σ_{al}, can be obtained when $n = 1$:

$$S_y = \sigma_{al} = \sqrt{\sigma_1^2 + \sigma_2^2 - \sigma_1\sigma_2} \tag{5.9}$$

where

$$\sigma_{1,2} = \frac{\sigma_x + \sigma_y}{2} \mp \sqrt{\left(\frac{\sigma_x - \sigma_y}{2}\right)^2 + \tau_{xy}^2} \tag{5.10}$$

Substituting Equation 5.9 into 5.10 yields

$$\sigma_{al} = \sqrt{\sigma_x^2 + 3\tau_{xy}^2} \tag{5.11}$$

Any load above σ_{al} will cause failure. According to this theory, the safety factor can also be written as

$$n = \frac{S_y}{\sqrt{\sigma_x^2 + 3\tau_{xy}^2}} \tag{5.12}$$

Example 5.1

A steel shaft with a diameter of 0.1 m is subjected to pure torque at a shearing stress of 62 MPa. Determine the safety factor if the material has a yield strength of 410 MPa psi.

SOLUTION

First, determine the principal stresses. Note that $\sigma_1 = -\sigma_2$ (see Figure 5.1).

$$\sigma_{1,2} = \frac{\sigma_x + \sigma_y}{2} \mp \sqrt{\left(\frac{\sigma_x - \sigma_y}{2}\right)^2 + \tau_{xy}^2}$$

Since $\sigma_x = 0$ and $\sigma_y = 0$, the above equation reduces to

$$\sigma_{1,2} = \mp\tau_{xy} = \pm 62 \text{ Mpa}$$

Using the distortion energy theory,

$$n = \frac{S_y}{\sqrt{\sigma_1^2 + \sigma_2^2 - \sigma_1\sigma_2}} = \frac{410}{\sqrt{62^2 + (-62)^2 + (62 \times 62)}} = 3.82$$

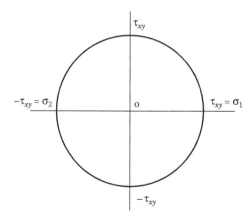

FIGURE 5.1 Principal stresses.

Example 5.2

A fixed steel bar shown in Figure 5.2 has a yield strength of 400 MPa. Determine whether a diameter of 0.03 m is safe.

SOLUTION

Calculate the maximum stress at point O due to external loads.

$$\sum F_x = 0 \Rightarrow -R_x + F_1 = 0, \quad R_x = 2\text{kN}$$

$$\sum F_y = 0 \Rightarrow R_y - F_2 = 0, \quad R_y = 2\text{kN}$$

$$\sum M_o = 0$$

$$M_x i + M_y j + M_z k + (\bar{r}_{OB} \times \bar{F}_2) + (\bar{r}_{OA} \times \bar{F}_1) = 0 \tag{a}$$

where

$$\bar{r}_{OB} = 0.6\bar{i} + 0.4\bar{k}$$

$$\bar{r}_{OA} = 0.6\bar{i}$$

$$\bar{F}_1 = 2\bar{i}$$

$$\bar{F}_2 = -2\bar{j}$$

Substituting into Equation (a) yields

$$M_x i + M_y j + M_z k = -[(0.6i + 0.4k) \times (-2j)] = +1.2k - 0.8i$$

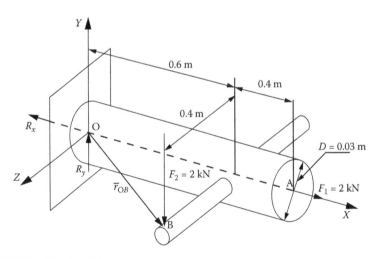

FIGURE 5.2 Fixed steel bar.

Note that the term $(\bar{r}_{OA} \times \bar{F}_1)$ goes to zero

$$M_x = -0.8\bar{i} \text{ kN-m}$$
$$M_z = -1.2\bar{k} \text{ kN-m}$$

Reaction at O is given by

$$M_x = -800 \text{ N m}$$
$$M_z = 1200 \text{ N m}$$
$$R_x = 2000 \text{ N}$$
$$R_y = 2000 \text{ N}$$

Figure 5.3 shows the following stresses at the critical element:

1. Stress due to $R_x = 2000$ N (tension):

$$\sigma_{x_1} = \frac{R_x}{A} = \frac{2000}{\pi/4(0.03)^2} = 2.83 \text{ MPa}$$

2. Stress due to M_z (tension):

$$\sigma_{x_2} = \frac{M_z C}{I} = \frac{1200 \times (0.03/2)}{\pi/64(0.03)^4} = 452.707 \text{ MPa}$$

3. Stress due to M_x (shear):

$$\tau_{xz} = \frac{Tr}{J} = \frac{M_x r}{J} = \frac{-800 \times (0.03/2)}{(\pi/32)(0.03)^4} = -150.9 \text{ MPa}$$

4. Total stress at the critical element (see Figure 5.4) is

$$\sigma_x = \sigma_{x_1} + \sigma_{x_2} = 2.83 + 452.707 = 455.54 \text{ MPa}$$

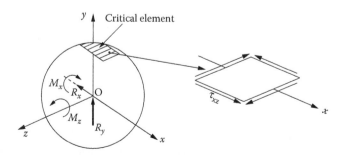

FIGURE 5.3   Critical stress element.

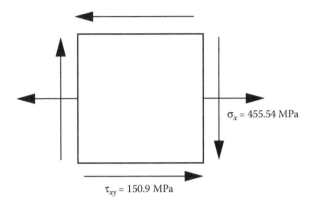

FIGURE 5.4 Total stress at critical stress element.

and shear stress is

$$\tau_{xz} = -150.9 \text{ MPa}$$

Using the distortion energy theory, the safety factor is

$$n = \frac{S_y}{\sqrt{\sigma_x + 3\tau_{xy}^2}} = \frac{400}{\sqrt{455.47^2 + 3 \times 150.9^2}} = 0.76$$

Since the safety factor is less than 1, a shaft with $D = 0.03$ m is not safe.

5.2 FATIGUE FAILURE

Although we have discussed that failure occurs when the stress level reaches the yield strength (for ductile material), the majority of structural member failures occur at stresses below the yield strength. This is not due to material defect but due to a phenomenon called fatigue. For flexural fatigue testing of metals, the R. R. Moore rotating beam testing system shown in Figure 5.5 is used to cyclically stress the specimen to fail by fracture. As shown in this figure, the rotating beam testing machine applies bending stress using dead weights as the specimen rotates. Standard test specimens shown in Figure 5.6 are tested at different loads to obtain data for plotting stress versus the number of cycles (S/N) in a curve. In general, the fatigue strength of materials is documented by the S–N curve obtained from constant amplitude imposed by a pure bending stress test.

The endurance strength, S_e, shown in Figure 5.7, has a safe range of fluctuating stress values, and below this value we assume that failure will not occur. The endurance limit implies that structural members stressed under this limit will have infinite life ($N_e = 10^6$ cycles). As shown in Figure 5.7, steel and titanium alloys (ferrous alloys such as low-strength carbon and alloy steel; some stainless steel, iron, and titanium alloys; and also some polymers) have an endurance limit and

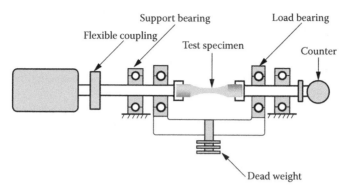

FIGURE 5.5 Rotating beam testing system.

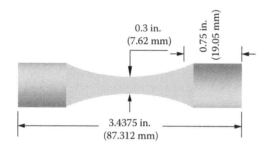

FIGURE 5.6 Standard fatigue test specimen.

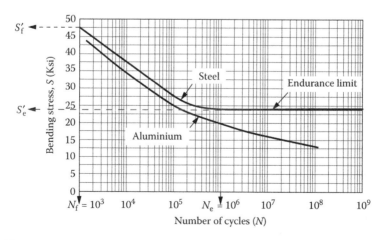

FIGURE 5.7 Typical S–N curve.

amplitude below which there appears to be no cycle that will cause failure. However, other structural metals such as aluminum, copper, magnesium, and nickel do not have an endurance limit and will eventually fail even from small stress amplitudes (see Figure 5.7). If the fatigue test data are not available, then the approximated S–N curve for a specific material can be constructed by using the values given in Table 5.1.

In Table 5.1, S'_e and S'_s are the uncorrected endurance limits for bending and shear, respectively. Once S'_e is known, endurance limits for shear can be determined by using the maximum shear theory or the Von Mises theory as shown in Table 5.1. Correction for S'_e will be discussed in the next section. Note that neither the tensile strength nor the yield strength is a good descriptor of the ductility or brittleness of a material.

S–N curves of materials are experimentally constructed by repeated pure bending, tension, or torsional loads. Data are presented in graphical form as shown in Figure 5.5 in which the fatigue life N in terms of cycles of load reversal is plotted against pure bending stress. If the S–N curve of a material is known, the endurance limit S_e can be found. The design of structural members can be accomplished by taking into consideration S_e so that the design provides infinite life.

If the S–N curve of the material used in design is not available, simple equations approximately representing the S–N curve can be used. The region beyond $N_f = 10^3$ cycles is called high-cycle fatigue. S_f corresponds to the number of cycles; $N_f = 10^3$ is called the fracture strength and can be estimated as $S'_f = 0.9 S_{ut}$ (for $S_{ut} \leq 250$ kpsi), as shown in Table 5.1. The equation of the S–N curve for steel shown in Figure 5.7 can be approximated by

$$S' = aN^b \tag{5.13}$$

TABLE 5.1
End Coordinates (N_f, S_f) and (N_e, S_e) of S–N Curve

Material	S'_f	N_f	S'_e	N_e	Endurance Limit in Shear, S_{se}	
					Maximum Shear	Von Mises
Ductile steel S_{ut} " 250 kpsi " (1724 MPa)	$0.9 S_{ut}$	10^3	$0.5 S_{ut}$	10^6	$0.5 S'_e$	$0.577 S'_e$
Hard steel $S_{ut} \geq 250$ kpsi $\geq$ (1724 MPa)	$0.9 S_{ut}$	10^3	$0.35 S_{ut}$	10^6	$0.5 S'_e$	$0.577 S'_e$
Copper alloys	$0.9 S_{ut}$	10^3	$(0.25–0.5)S_{ut}$	10^6	$0.5 S'_e$	$0.577 S'_e$
Nickel alloys	$0.9 S_{ut}$	10^3	$(0.35–0.5)S_{ut}$	10^6	$0.5 S'_e$	$0.577 S'_e$
Aluminum alloys	$0.9 S_{ut}$	10^3	$0.35 S_{ut}$	10^6	$0.5 S'_e$	$0.577 S'_e$
Magnesium alloys	$0.9 S_{ut}$	10^3	$0.35 S_{ut}$	10^6	$0.5 S'_e$	$0.577 S'_e$
Titanium	$0.9 S_{ut}$	10^3	$(0.45–0.65)S_{ut}$	10^6	$0.5 S'_e$	$0.577 S'_e$

where

$$b = \frac{\log(S_f'/S_e')}{\log(N_f/N_e)} \tag{5.14}$$

and

$$a = \frac{S_f'^2}{S_e'} \tag{5.15}$$

If the stress, σ, in structural members falls between the fracture strength, S_f, and the endurance strength, S_e, fatigue life can be determined from Equation 5.13 as

$$N = N_f \left(\frac{\sigma}{S_f'}\right)^{1/b} \tag{5.16}$$

Example 5.3

Construct an approximated S–N curve for a steel bar, assuming that $S_{ut} = 500$ MPa. Determine the number of cycles, if the stress in the steel bar is 350 MPa.

SOLUTION

Using Table 5.1, the uncorrected endurance limit is

$$S_e' = 0.5\,S_{ut} = 0.5(500) = 250\,\text{MPa} \quad \text{at } N_e = 10^6 \text{ cycles}$$

and the uncorrected fracture strength for $N_f = 10^3$ cycles is

$$S_f' = 0.9\,S_{ut} = 0.9(500) = 450\,\text{MPa}$$

Using calculated values, a simple approximated S–N curve shown in Figure 5.8 can be constructed. From the S–N curve, the number of cycles corresponding to 350 MPa is determined to be 17×10^3 cycles.

Example 5.4

If the ultimate strength of a steel bar is 400 MPa, determine the fatigue strength corresponding to a life of 90×10^3 cycles.

SOLUTION

Estimate the endurance limit and fatigue fracture strength using Table 5.1

$$S_e' = 0.5\,S_{ut} = 0.5 \times 400 = 200\,\text{MPa}$$

$$S_f' = 0.9\,S_{ut} = 0.9 \times 400 = 360\,\text{MPa}$$

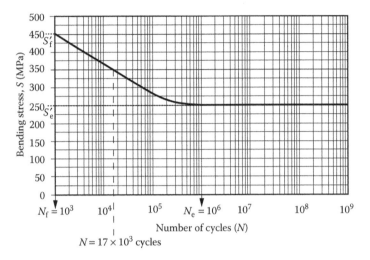

FIGURE 5.8 S–N curve.

From Table 5.1, S_e starts at 10^6 cycles and S_f corresponds to 10^3 cycles. Hence, the constants a and b are given as

$$a = \frac{S_f'^2}{S_e'} = \frac{360^2}{200} = 648$$

$$b = \frac{\log(S_f'/S_e')}{\log(N_f/N_e)} = \frac{\log(360/200)}{\log(10^3/10^6)} = -0.085$$

Thus, the finite life strength corresponding to 90×10^3 cycles is

$$S' = aN^b = 648(90 \times 10^3)^{-0.085} = 245.7 \text{ MPa}$$

5.2.1 FATIGUE STRENGTH CORRECTION FACTORS

Since the rotating beam specimen used in the laboratory to determine the fatigue strength is different in size, surface finish, and operating environment, the endurance limit of the machine components in consideration should be modified. In general, the fatigue strength of machine components is lower than the one found using laboratory testing [1,2].

Using appropriate modifying factors, the corrected endurance strength is defined as

$$S_e = S_e' c_s c_z c_t c_r c_f c_m \qquad (5.17)$$

where S_e is the corrected endurance strength of machine components, S_e is the endurance strength of the test specimen, c_s is the surface finish correction factor, c_z is the size correction factor, c_t is the temperature correction factor, c_r is the reliability correction factor, c_f is the fatigue stress concentration correction factor, and c_m is the miscellaneous correction factor.

5.2.1.1 Surface Finish Correction Factor

Fatigue strength is sensitive to surface irregularities. To eliminate the effect of surface irregularities, the surface of the rotating beam specimen is highly polished. Admittedly, most machine components do not have such a high-quality surface finish. When the ultimate strength in tension for steel is known, the surface correction factor can be obtained from Figure 5.9.

5.2.1.2 Size Correction Factor

Assume that the diameter of a standard test specimen used to obtain fatigue strength S_e is 0.3 in. (7.62 mm). When the machine component is larger than the standard specimen dimension, it demonstrates less strength, and so S_e is adjusted to reflect the reduction in strength. Shigley and Mitchell [2] proposed the following expression

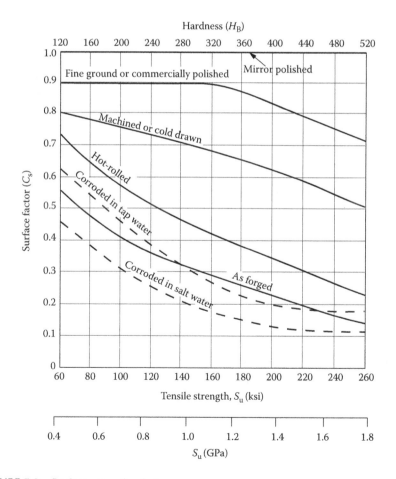

FIGURE 5.9 Surface correction factors.

for rotating round cylindrical machine components when they are under bending and torsion stress:

$$c_z = \left|\left(\frac{d}{0.3}\right)^{-0.1133}\right. \quad \text{for } 0.11 \text{ in} \le d \le 2 \text{ in} \quad \left|\left(\frac{d}{7.62}\right)^{-0.1133}\right. \quad \text{for } 2.79\,\text{mm} \le d \le 51\,\text{mm}$$

(5.18)

For larger size components, the size correction factor, c_z, varies from 0.60 to 0.75 for bending and torsion. Since there is no size effect for axial loading, $c_z = 1$. The effect of size is mainly important for the analysis of rotating shafts. For situations in which machine components do not rotate or do not have a round cross section, the equivalent diameter, d_{eq}, is used in Equation 5.18. This is obtained by equating 95% of the stress area of the rotating round shaft to the same stress area of the component under consideration [2].

The equivalent diameter, d_{eq}, for a nonrotating round cross section under bending is given as

$$d_{eq} = 0.37d$$

(5.19)

The equivalent diameter, d_{eq}, for a rectangular section under bending with width w and thickness t is given as

$$d_{eq}^2 = 0.65\,wt$$

(5.20)

5.2.1.3 Temperature Correction Factor

The endurance limit, S_e, is typically determined from the standard rotating beam test machine at room temperature. Since higher temperatures cause a reduction in strength, the temperature correction factor to modify the endurance limit of steel given in Equation 5.21 can be used for reasonably high temperatures [2].

$$
\begin{aligned}
&\text{For } T \le 450°\text{C } (840°\text{F}): && c_t = 1 \\
&\text{For } 450°\text{C} \le T \le 550°\text{C}: && c_t = 1 - 0.0058(T - 450) \\
&\text{For } 840°\text{F} \le T \le 1020°\text{F}: && c_t = 1 - 0.0032(T - 840)
\end{aligned}
$$

(5.21)

5.2.1.4 Reliability Correction Factor

The S–N curve values are mean values based on a number of tests resulting in loads at failure. The endurance strength values are mean values implying a 50% survival rate. To determine the endurance strength with a survival rate higher than 50%, the endurance strength values must be reduced. Assuming an 8% standard deviation for both stress and endurance strength, the reliability correction factors, c_r, shown in Table 5.2 are used.

TABLE 5.2
Reliability Correction Factors

Reliability (%)	c_r
50	1
90	0.897
95	0.868
99	0.814
99.9	0.753
99.99	0.702
99.999	0.659
99.9999	0.620

5.2.1.5 Fatigue Stress Concentration Correction Factor

In general, stress analysis calculations assume that the machine components have uniform sections without any irregularities or discontinuities. However, many machine components and parts have discontinuities such as holes, grooves, fillets, shoulders, keyways, screws, threads, and notches that affect the stress distribution near these discontinuities and cause local stress to increase. As shown in Figure 5.10, this increase in stress is called "stress concentration." As shown in this figure, stress distribution becomes uniform away from the discontinuities.

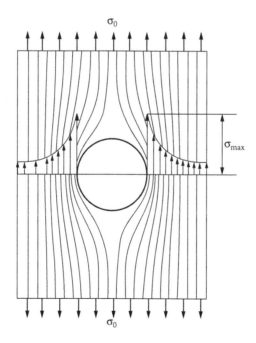

FIGURE 5.10 Stress concentration.

The theoretical stress concentration factors K_t, used for normal stress, and K_{ts}, used for shear stress, are defined as

$$K_t = \frac{\sigma_{max}}{\sigma_0} \qquad (5.22)$$

$$K_{ts} = \frac{\tau_{max}}{\tau_0} \qquad (5.23)$$

The value of a stress concentration factor is a function of the geometry, the type of discontinuity, and the type of load applied on the machine components. Stress concentration factors for different geometries and load types are shown in Figures A2.1 through A2.5 (see Appendix 2).

Once the theoretical stress concentration factor is known, the fatigue stress concentration correction factor, c_f, can be determined as

$$c_f = \frac{1}{k_f} \qquad (5.24)$$

where the fatigue strength reduction factor, k_f, is given as

$$k_f = 1 + q(K_t - 1) \qquad (5.25)$$

where q is the notch sensitivity factor. Figure 5.11 shows the value of notch sensitivity factors for steel and aluminum.

5.2.1.6 Miscellaneous Correction Factor

The miscellaneous correction factor, c_m, is a general factor to allow for different factors that are not easily quantifiable. These factors may include the effect of impact, corrosion, environment, and metal spraying among others. Therefore, the endurance limit, S_e, determined in the laboratory must be modified by c_m.

Example 5.5

The fine ground rotating shaft shown in Figure 5.12 is made of steel that has an ultimate strength of $S_{ut} = 700$ MPa. Assume that the expected reliability is 99% and the load applied is bending. Determine the fatigue strength if the machine operating temperature is 80°C.

SOLUTION

The uncorrected fatigue strength is

$$S'_e = 0.5\, S_{ut} = 0.5 \times 700 = 350 \text{ MPa}$$

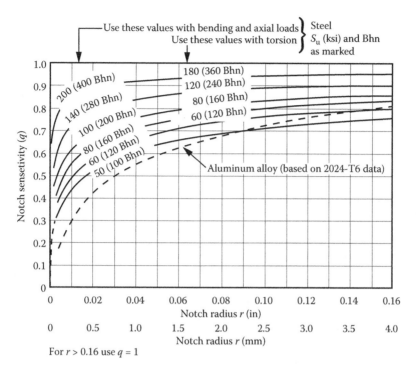

FIGURE 5.11   Notch sensitivity factors for steel and aluminum.

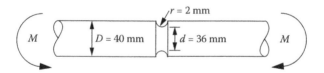

FIGURE 5.12   Rotating shaft under bending stress.

From Figure 5.9, the surface correction factor is found to be $c_s = 0.9$. The size correction factor, c_z, from Equation 5.18 is

$$c_z = \left(\frac{d}{7.62}\right)^{-0.1133} = \left(\frac{40}{7.62}\right)^{-0.1133} = 0.83$$

From Equation 5.21, the temperature correction factor is $c_t = 1$. From Table 5.2, the reliability correction factor is $c_r = 0.814$. Using Figure A2.3 (see Appendix 2), we calculate the ratios for stress concentration factor, K_t, as

$$\frac{r}{d} = \frac{2}{36} = 0.055$$

and

$$\frac{D}{d} = \frac{40}{36} = 1.11$$

and find $K_t = 2.25$. Using $S_{ut} = 700$ MPa (101.5 ksi) from Figure 5.11, the notch sensitivity factor is $q = 0.82$. The fatigue stress concentration correction factor, c_f, can be determined as

$$c_f = \frac{1}{1 + q(K_t - 1)} = \frac{1}{1 + 0.82(2.25 - 1)} = 0.49$$

Then the corrected endurance limit is

$$S_e = S'_e C_s C_z C_t C_r C_f = 350 \times 0.9 \times 0.83 \times 1 \times 0.814 \times 0.49 = 104.3 \text{ MPa}$$

5.3 FLUCTUATING STRESSES

So far, fatigue analysis has been explained based on a fully reversed load as shown in Figure 5.13. In some cases, stress applied on structural members fluctuates without passing through zero. For example, two unique cases are shown in Figure 5.14. In this case (where the applied force does not cause complete reversal), the combination of stress amplitude σ_a and mean stress σ_m must be considered in fatigue analysis. Hence, the effect of mean stress is included in the calculation of fatigue life.

Two approaches will be used to determine the safety factor and the equivalent bending stress, which takes into account the combination of stress amplitude σ_a and mean stress σ_m.

5.3.1 FATIGUE ANALYSIS FOR BRITTLE MATERIALS

The modified Goodman diagram shown in Figure 5.15 is used for the fatigue analysis of brittle materials. Any design on the lines of the Goodman diagram will be safe, with a safety factor of $n = 1$. In the case of brittle materials, the stress concentration factor, K_t, increases the likelihood of failure. Therefore, the stress concentration factor will be used for both stress amplitude σ_a and mean stress σ_m. From Figure 5.15, we have the following relationship:

$$\frac{GD}{OG} = \frac{EG}{OC} \tag{5.26}$$

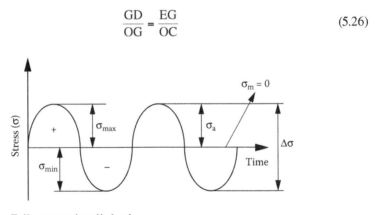

FIGURE 5.13 Fully reversed cyclic load.

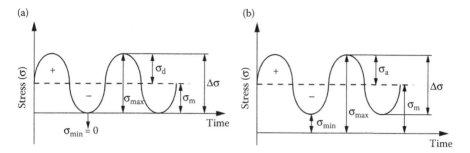

FIGURE 5.14 (a) Repeated cyclic load and (b) fluctuating cyclic load.

where

$$GD = \frac{SE}{n} - K_t\sigma_m$$
$$EG = K_t\sigma_a$$
$$OD = \frac{S_{ut}}{n} \qquad (5.27)$$
$$OC = \frac{S_e}{n}$$

Substituting Equation 5.27 into Equation 5.26 yields the safety factor for brittle materials in bending:

$$n_{brittle} = \frac{S_{ut}}{K_t\left(\sigma_m + (S_{ut}/S_e)\sigma_a\right)} \qquad (5.28)$$

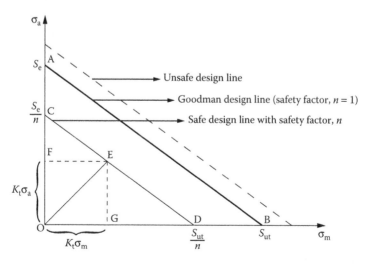

FIGURE 5.15 Modified Goodman diagram used for brittle material fatigue analysis.

If the loading is only fully reversed cyclic, that is, if the steady stress component σ_m in Equation 5.28 does not exist, then the safety factor in bending reduces to

$$n_{\text{brittle}} = \frac{S_e}{k_t \sigma_a} \tag{5.29}$$

Equation 5.29 can be used to define the safety factor in shear as

$$n_{\text{brittle}} = \frac{S_{se}}{k_{ts} \tau_a} \tag{5.30}$$

where S_{se} is the endurance limit in shear and τ_a is the shear stress amplitude. In Equation 5.28 when $K_t = 1$, we obtain the following modified Goodman equation:

$$\frac{\sigma_a}{S_e} + \frac{\sigma_m}{S_{ut}} = \frac{1}{n_{\text{brittle}}} \tag{5.31}$$

If we assume that the design is on the lines of the Goodman diagram, further simplification can help to determine the equivalent bending stress amplitude σ_{eq}, which takes into account the combination of stress amplitude σ_a and mean stress σ_m.

$$\sigma_{eq} = \sigma_a \left(\frac{S_{ut}}{S_{ut} - \sigma_m} \right) \tag{5.32}$$

Note that in Equation 5.28, S_e is replaced by σ_{eq}. The increasing effect of the mean stress due to tension on σ_a is evident from Equation 5.32. However, there is a reduction in equivalent bending stress amplitude with increasing tensile mean stress. Experiments show that compressive stress has no effect on stress amplitude, σ_a. In other words, if mean stress is due to compressive stress, Equation 5.32 should not be used. As shown in Figures 5.14a and b, since the repeated cyclic load and fluctuating cyclic load have a mean stress, Equation 5.32 should be used to calculate fatigue life, provided that the mean stress σ_m is due to tensile loading.

If σ_{min} and σ_{max} are known, σ_a, σ_m, and σ_r (stress range) can be determined as

$$\sigma_a = \frac{\sigma_{max} - \sigma_{min}}{2}$$

$$\sigma_m = \frac{\sigma_{max} + \sigma_{min}}{2} \tag{5.33}$$

$$\sigma_r = 2\sigma_a$$

Example 5.6

As shown in Figure 5.16, one end of the fixed solid steel bar is subjected to a constant tensile load of $F = 8$ kN and a cyclic load of $P = 2$ kN.

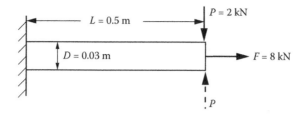

FIGURE 5.16 Solid steel bar under combined loading.

a. Assuming that ultimate strength of the steel bar is 700 MPa, determine the resulting equivalent stress amplitude on the steel bar. Ignore the stress concentration effect.

b. If the tensile load applied on the steel bar is compressive load, determine the equivalent stress.

SOLUTION

a. Since the solid bar is subjected to a constant mean stress due to the applied tensile load of 8 kN, Equation 5.32 should be used.

$$\sigma_{eq} = \sigma_a \left(\frac{S_{ut}}{S_{ut} - \sigma_m} \right)$$

where mean stress σ_m is

$$\sigma_m = \frac{F}{A}$$

where

$$A = \frac{\pi D^2}{4} = \frac{\pi (0.03)^2}{4} = 706.5 \times 10^{-6} \, m^2$$

Then

$$\sigma_m = \frac{F}{A} = \frac{8000}{706.5 \times 10^{-6}} = 11.32 \times 10^6 \, Pa$$

The stress amplitude due to the cyclic bending load is

$$\sigma_a = \frac{Mc}{I}$$

where

$$c = \frac{D}{2} = \frac{0.03}{2} = 0.015\text{m}$$

$$M = PL = 2000 \times 0.5 = 1000 \text{ Nm}$$

$$I = \frac{\pi D^4}{64} = \frac{\pi(0.03)^4}{64} = 3.98 \times 10^{-8} \text{ m}^4$$

$$\sigma_a = \frac{Mc}{I} = \frac{(1000)(0.015)}{3.98 \times 10^{-8}} = 3.77 \times 10^8 \text{ Pa}$$

The equivalent stress, σ_{eq}, which takes into account the combination of stress amplitude, σ_a, and mean stress, σ_m, is

$$\sigma_{eq} = \sigma_a \left(\frac{S_{ut}}{S_{ut} - \sigma_m} \right)$$

Substituting known values yields

$$\sigma_{eq} = 377 \left(\frac{700}{700 - 11.32} \right) = 383.2 \text{ MPa}$$

b. Since the mean stress is due to compressive load

$$\sigma_{eq} = \sigma_a = 377 \text{ MPa}$$

Compressive load has no effect on σ_a.

5.3.2 FATIGUE ANALYSIS FOR DUCTILE MATERIALS

In this case, as shown in Figure 5.17, the Soderberg diagram is used for fatigue analysis. Note that since the material is ductile, S_{ut} is replaced by S_y.

The same analogy is used to obtain the safety factor for ductile materials.

$$n_{\text{ductile}} = \frac{S_y}{\sigma_m + (S_y/S_e)k_f\sigma_a} \tag{5.34}$$

As seen from Figure 5.17 and Equation 5.34, only the alternating stress component, σ_a, is affected by the fatigue strength reduction factor, k_f, not the steady stress component, σ_m. If the loading is only fully reversed cyclic, the safety factor in bending for ductile materials reduces to

$$n_{\text{ductile}} = \frac{S_e}{k_f\sigma_a} \tag{5.35}$$

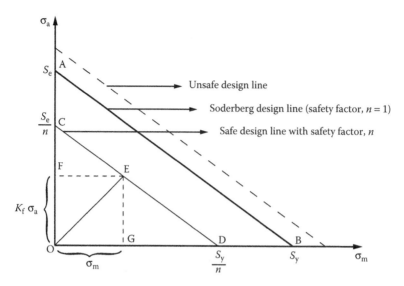

FIGURE 5.17   Soderberg diagram used for ductile material fatigue analysis.

and the safety factor in shear for ductile materials to

$$n_{\text{ductile}} = \frac{S_{se}}{k_{fs}\tau_a} \tag{5.36}$$

Example 5.7

As shown in Figure 5.18, a shaft with a diameter of 40 mm is subjected to a fluctuating bending stress with $\sigma_{\text{max}} = 100$ MPa and $\sigma_{\text{min}} = -10$ MPa. The shaft is machined and has $S_{ut} = 800$ MPa and $S_y = 400$ MPa. Determine the safety factor for fatigue failure if 95% reliability is desired. Assume that the material is ductile.

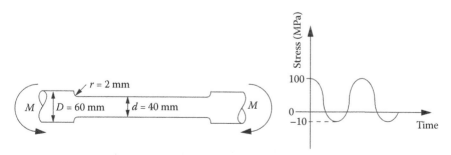

FIGURE 5.18   Rotating shaft under fluctuating bending stress.

SOLUTION

The uncorrected fatigue strength is

$$S'_e = 0.5\,S_{ut} = 0.5 \times 800 = 400\ \text{MPa}$$

From Figure 5.9, the surface correction factor is $c_s = 0.74$. The size correction factor, c_z, from Equation 5.18 is

$$c_z = \left(\frac{d}{7.62}\right)^{-0.1133} = \left(\frac{40}{7.62}\right)^{-0.1133} = 0.83$$

Since the operational temperature is not mentioned, the temperature correction factor is $c_t = 1$. From Table 5.2, the reliability correction factor is $c_r = 0.868$. Using Figure A2.1 (see Appendix 2), we calculate the ratios for the stress concentration factor, K_t, as

$$\frac{r}{d} = \frac{2}{40} = 0.05$$

$$\frac{D}{d} = \frac{60}{40} = 1.5$$

and find $K_t = 2.03$. Using $S_{ut} = 800\ \text{MPa}$ (116 ksi) from Figure 5.11, the notch sensitivity factor is $q = 0.83$. The fatigue strength reduction factor, k_f, can be determined as

$$k_f = 1 + q(K_t - 1) = 1 + 0.83(2.03 - 1) = 1.85$$

and the fatigue stress concentration correction factor is

$$c_f = \frac{1}{1.85} = 0.54$$

Then, the corrected endurance limit is

$$SE = S'_e c_s c_z c_t c_r c_f = 400 \times 0.74 \times 0.83 \times 1 \times 0.868 \times 0.54 = 115.2\ \text{MPa}$$

Stress amplitude is

$$\sigma_a = \frac{\sigma_{max} - \sigma_{min}}{2} = \frac{100 - (-10)}{2} = 55\ \text{MPa}$$

Mean stress is

$$\sigma_m = \frac{\sigma_{max} + \sigma_{min}}{2} = \frac{100 + (-10)}{2} = 45\ \text{MPa}$$

Then the safety factor for ductile materials is

$$n_{ductile} = \frac{S_y}{\sigma_m + (S_y/S_e)k_f \sigma_a} = \frac{400}{45 + (400/115.2) \times 1.85 \times 55} = 1$$

The safety factor is right on the line of the diagram. Design requirements must be changed to have the safety factor at least above 1.5.

5.4 CUMULATIVE FATIGUE DAMAGE

Fatigue life under variable amplitude load is estimated by the linear damage rule, also known as Miner's rule. Miner's rule was first proposed by Palmgren in 1924 and was further developed by Miner in 1945. Miner's rule is defined as follows:

$$\sum_{i=1}^{I} \left(\frac{n_i}{N_i} \right) = C^* \tag{5.37}$$

where n is the number of cycles of stress amplitude σ_a applied to the machine component and N is the life corresponding to that stress amplitude. C^* is experimentally found to be between 0.7 and 2.2. Usually for design purposes, C^* is assumed to be 1, which gives a gross estimate of fatigue life over the load spectrum.

Miner's rule is used to evaluate the cumulative damage of mechanical components at stress levels exceeding the endurance limit. According to Miner's rule, if a mechanical component is subjected to a constant amplitude-alternating load and fails after N_f cycles, each cycle expends $1/N_f$ fraction of the component's total life. This can be thought of as determining what proportion of total life is consumed by stress reversal at each magnitude and then forming a linear combination of their damages. For example, a machine component is subjected to n_1 reversed stress cycle at a stress level of σ_1, n_2 reversed stress cycle at a stress level of σ_2, and so on. As shown in Figure 5.19, assume that the number of cycles to failure at stress level σ_1 is N_1 cycles, σ_2 is N_2 cycles, and so on. By Miner's rule, the machine component failure is predicted to occur when

$$\frac{n_1}{N_1} + \frac{n_2}{N_2} + \cdots + \frac{n_i}{N_i} \geq 1 \tag{5.38}$$

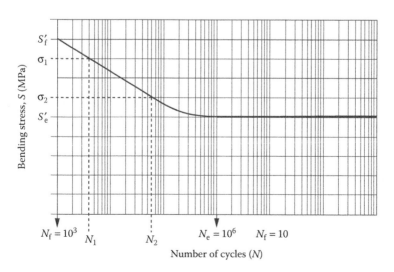

FIGURE 5.19 Fatigue life prediction by Miner's rule.

Example 5.8

As shown in Figure 5.20, a solid steel bar with one end fixed is first subjected to a cyclic load of 20,000 lb for 6000 cycles, then the magnitude of the cyclic load is reduced to 3000 lb for 50,000 cycles, and finally the cyclic load is increased to 10,000 lb for 1000 cycles. Using the S–N curve given in Figure 5.21, calculate the remaining fatigue life. Assume $S_{ut} = 100$ kpsi, $S_y = 80$ kpsi, $E = 30 \times 10^6$ psi, and ignore the fatigue correction factors.

SOLUTION

The area A and the moment of inertia I of the steel bar are

$$A = \frac{\pi D^2}{4} = \frac{\pi \times 5^2}{4} = 19.63 \, \text{in.}^2$$

$$I = \frac{\pi D^4}{64} = \frac{\pi \times 5^4}{64} = 30.67 \, \text{in.}^4$$

Maximum stress for a load of 20,000 lb is

$$\sigma_{max_1} = \frac{Mc}{I} = \frac{20,000 \times 25 \times 5/2}{30.67} = 40,756.4 \, \text{psi}$$

Thus, the mean stress and stress amplitude can be given by

$$\sigma_{a_1} = \sigma_{m_1} = \frac{\sigma_{max_1}}{2} = \frac{40,756.4}{2} = 20,378.2 \, \text{psi} \, (20.378 \, \text{kpsi})$$

The equivalent stress is

$$\sigma_{eq_1} = \sigma_{a_1}\left(\frac{S_{ut}}{S_{ut} - \sigma_m}\right) = 20.378\left(\frac{100}{100 - 20.378}\right) = 25.6 \, \text{kpsi}$$

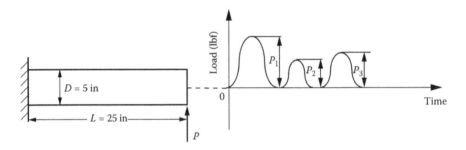

FIGURE 5.20   Solid cantilever bar under cyclic loading.

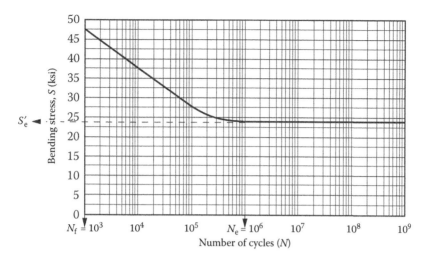

FIGURE 5.21 S–N curve for solid bar material.

From the S–N curve given in Figure 5.21, the life cycle corresponding to a stress level of 25.6 kpsi is

$$N_1 = 200{,}000 \text{ cycles}$$

The maximum stress for a load of 3000 lb is

$$\sigma_{max_2} = \frac{Mc}{I} = \frac{3000 \times 25 \times 5/2}{30.67} = 6113 \text{ psi}$$

Thus, the mean stress and stress amplitude can be given by

$$\sigma_{a_2} = \sigma_{m_2} = \frac{\sigma_{max_2}}{2} = \frac{6.113}{2} = 3.057 \text{ kpsi}$$

The equivalent stress is

$$\sigma_{eq_2} = \sigma_{a_2}\left(\frac{S_{ut}}{S_{ut} - \sigma_m}\right) = 3.057\left(\frac{100}{100 - 3.057}\right) = 3.153 \text{ kpsi}$$

From the S–N curve given in Figure 5.21, the life cycle corresponding to a stress level of 3.149 kpsi is infinite ($N_2 = \infty$).
The maximum stress for a load of 10,000 lb is

$$\sigma_{max_3} = \frac{Mc}{I} = \frac{10{,}000 \times 25 \times 5/2}{30.67} = 20.378 \text{ psi}$$

Thus, the mean stress and stress amplitude can be given by

$$\sigma_{a_3} = \sigma_{m_3} = \frac{\sigma_{max_3}}{2} = \frac{20.378}{2} = 10.20 \text{ kpsi}$$

The equivalent stress is

$$\sigma_{eq_3} = \sigma_{a_3}\left(\frac{S_{ut}}{S_{ut} - \sigma_m}\right) = 10.20\left(\frac{100}{100 - 10.20}\right) = 11.36\,\text{kpsi}$$

From the S–N curve given in Figure 5.21, the life cycle corresponding to a stress level of 11.36 kpsi is infinite ($N_3 = \infty$). Apply Miner's rule to determine the percentage damage as

$$\text{Total damage} = \sum\left(\frac{n_i}{N_i}\right)100 = \left(\frac{n_1}{N_1} + \frac{n_2}{N_2} + \frac{n_3}{N_3}\right) \times 100$$

$$= \left(\frac{6000}{200,000} + \frac{50,000}{\infty} + \frac{1000}{\infty}\right) \times 100 = (0.03 + 0 + 0) \times 100$$

$$= 3\%$$

Then the remaining life = $100 - 3 = 97\%$

5.5   DESIGN ANALYSIS USING FRACTURE MECHANICS

Fracture mechanics is the field of mechanics related to the study of the formation of cracks in materials. In general, failures by fracture occur under loads below the yield strength of the material. Three possible material separation modes for crack extension under external load are shown in Figure 5.22. In most of engineering practices, mode II, in-plane shear, and mode III, out-of-plane shear, have limited application. Therefore, only mode I, tensile opening mode, will be discussed in the following section.

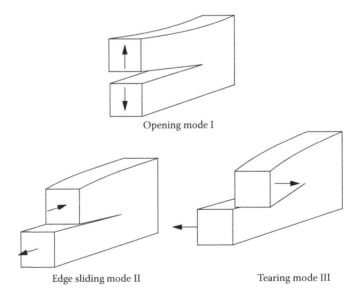

Opening mode I

Edge sliding mode II   Tearing mode III

FIGURE 5.22   Fracture modes.

5.5.1 STRESS STATE IN A CRACK (MODE I)

Linear elastic fracture mechanics can be used to provide preliminary design guide-lines. As shown in Figure 5.23, a plate is subjected to tensile stress σ_y at infinity with a crack length of $2a$. An element $dxdy$ of the plate at a distance r from the crack tip and at an angle θ with respect to the crack plane will have the following stress field [3]:

$$\sigma_x = \frac{K_I}{\sqrt{2\pi r}}\cos\frac{\theta}{2}\left(1 - \sin\frac{\theta}{2}\sin\frac{3\theta}{2}\right) \tag{5.39}$$

$$\sigma_y = \frac{K_I}{\sqrt{2\pi r}}\cos\frac{\theta}{2}\left(1 + \sin\frac{\theta}{2}\sin\frac{3\theta}{2}\right) \tag{5.40}$$

$$\tau_{xy} = \frac{K_I}{\sqrt{2\pi r}}\left(\sin\frac{\theta}{2}\cos\frac{\theta}{2}\cos\frac{3\theta}{2}\right) \tag{5.41}$$

$$\sigma_z = \nu\left(\sigma_x + \sigma_y\right), \ \tau_{xy} = \tau_{yz} = 0 \quad \text{for plane strain} \tag{5.42}$$

$$\sigma_z = 0 \text{ for plane stress} \tag{5.43}$$

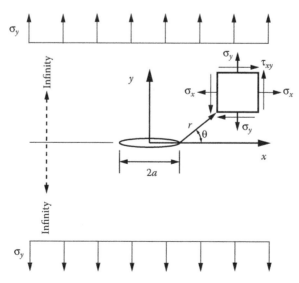

FIGURE 5.23 Crack in an infinite plate subjected to a tensile stress.

where K_I is the stress intensity factor. The magnitude of K_I depends on load (in this case, mode I, tensile load), structural geometry, size, and location of the crack. The expression for K_I is given as

$$K_I = \sigma\sqrt{\pi a} \qquad\qquad (5.44)$$

Using correction factor $f(a/b)$, Equation 5.44 can be modified to determine the stress intensity factor of the crack type given in Figure 5.24.

When the normal stress σ is equal to S_y, the material becomes unstable and plastic deformation occurs. Note that a similar analogy can be used for material failure such that when the stress intensity factor K_I reaches the critical stress intensity factor K_{IC},

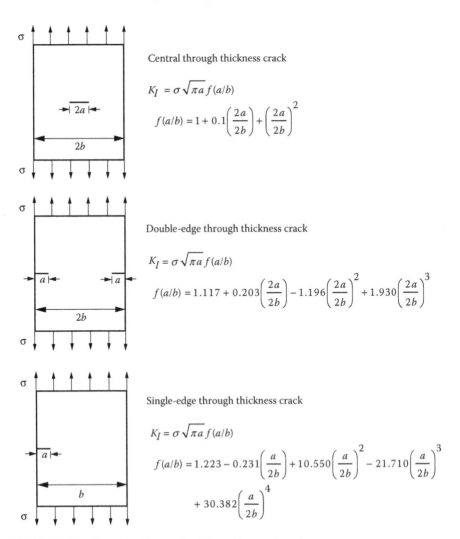

Central through thickness crack

$$K_I = \sigma\sqrt{\pi a}\, f(a/b)$$

$$f(a/b) = 1 + 0.1\left(\frac{2a}{2b}\right) + \left(\frac{2a}{2b}\right)^2$$

Double-edge through thickness crack

$$K_I = \sigma\sqrt{\pi a}\, f(a/b)$$

$$f(a/b) = 1.117 + 0.203\left(\frac{2a}{2b}\right) - 1.196\left(\frac{2a}{2b}\right)^2 + 1.930\left(\frac{2a}{2b}\right)^3$$

Single-edge through thickness crack

$$K_I = \sigma\sqrt{\pi a}\, f(a/b)$$

$$f(a/b) = 1.223 - 0.231\left(\frac{a}{2b}\right) + 10.550\left(\frac{a}{2b}\right)^2 - 21.710\left(\frac{a}{2b}\right)^3$$

$$+ 30.382\left(\frac{a}{2b}\right)^4$$

FIGURE 5.24   Correction factors for different types of cracks.

significant crack propagation occurs. Thus, the design engineer must keep the K_I value lower than the K_{IC} value in the same way that the normal stress, σ, due to the applied force must be lower than the yield strength, S_y, for a safe design. Once the value of K_{IC} for a material of a particular thickness is known, the design engineer can determine the crack size that can be allowed in structural members for a given stress level.

Example 5.9

Figure 5.25 shows a high-grade steel plate used in ship building. Assume that $a = 40$ mm through-thickness crack is subjected to a tensile stress. The steel plate has a critical stress intensity factor $K_{IC} = 30\,\text{MPa}\sqrt{m}$. Determine the maximum tensile stress for failure.

SOLUTION

Since the steel plates used in the ship-building industry are long and wide, we assume that a through-thickness crack is at the center of the plate. Then, the infinite plate formula as in Equation 5.44 can be modified to

$$K_{IC} = \sigma_{max}\sqrt{\pi a}$$

where σ_{max} is the maximum stress for failure and is given as

$$\sigma_{max} = \frac{K_{IC}}{\sqrt{\pi a}} = \frac{30\sqrt{10^3}}{\sqrt{\pi \times 40}} = 84.65\,\text{MPa}$$

5.5.2 ELLIPTICAL CRACK IN AN INFINITE PLATE

The problem of an imbedded elliptical or semielliptical crack in an infinite solid has attracted much attention in engineering application. In pressurized components and vessels such as pressure vessels and pipelines, a crack easily develops from small defects and material imperfections.

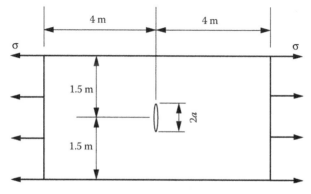

FIGURE 5.25 Through-thickness crack in a steel plate.

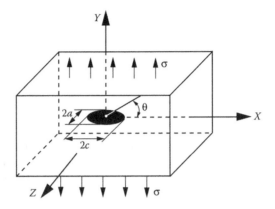

FIGURE 5.26 Elliptical crack in an infinite solid.

The commonly used approximation for the stress intensity factor K_I at any point along the perimeters of elliptical or circular cracks embedded in an infinite solid body subjected to uniform uniaxial tension (see Figure 5.26) is defined by [4]

$$K_I = \frac{\sigma\sqrt{\pi a}}{\Phi}\left(\sin^2\theta + \frac{a^2}{c^2}\cos^2\theta\right)^{1/4} \tag{5.45}$$

where Φ is the elliptical integral and is defined as

$$\Phi = \int_0^{\pi/2}\left(\sqrt{1 - k^2\sin^2\varphi}\right)d\varphi \tag{5.46}$$

where

$$k^2 = 1 - \left(\frac{a}{c}\right)^2 \tag{5.47}$$

The elliptic integral Φ has the following series expansions:

$$\Phi = \frac{\pi}{2}\left[1 - \frac{1}{4}\left(\frac{c^2 - a^2}{c^2}\right) - \frac{3}{64}\left(\frac{c^2 - a^2}{c^2}\right)^2 - \cdots\right] \tag{5.48}$$

Equation 5.48 can be approximated by neglecting higher-order terms

$$\Phi = \frac{3\pi}{8} + \frac{\pi}{8}\left(\frac{a}{c}\right)^2 \tag{5.49}$$

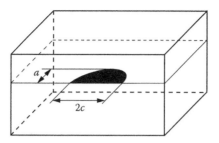

FIGURE 5.27 Semielliptical surface crack.

Equation 5.45 can be modified by including the back-crack free correction factor of 1.1 and the effective crack length a^*, which takes into account the plastic zone at the crack tip to determine the stress intensity factor for the semielliptical surface crack shown in Figure 5.27.

$$K_I = 1.1 \frac{\sigma}{\Phi} \sqrt{\pi a^*} \left(\sin^2 \theta + \frac{a^2}{c^2} \cos^2 \theta \right)^{1/4} \tag{5.50}$$

where

$$a^* = a + \frac{K_I^2}{4\pi \sqrt{2} S_y^2}$$

Including the effect of the crack shape around the crack front, Equation 5.50 can be further modified to determine the maximum value of the stress intensity factor at the minor axis ($\theta = \pi/2$) as follows:

$$K_I = 1.1\sigma \sqrt{\pi \frac{a}{Q}} \tag{5.51}$$

where Q is the crack-shape parameter given as

$$Q = \Phi^2 - 0.212 \left(\frac{\sigma}{S_y} \right)^2 \tag{5.52}$$

Finally, a magnification factor M_k for deep cracks can be used to find maximum stress intensity factor for a semielliptical surface flow.

$$K_I = 1.1 M_k \sigma \sqrt{\pi \frac{a}{Q}} \tag{5.53}$$

As shown in Figure 5.28, M_k values can be assumed to vary linearly as a/t varies from 0.5 to 1.0. M_k value is assumed to be unity for a/t values less than 0.5 [6].

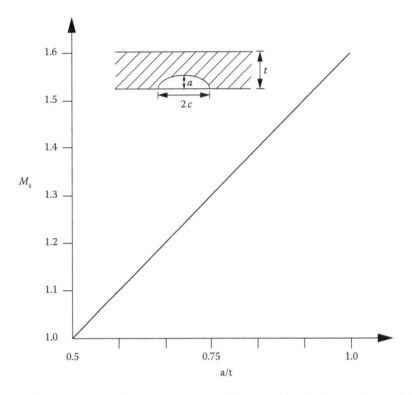

FIGURE 5.28 Magnification factor, MK. (Adapted from Rolfe, S. T. and Barsom, J. M. 1977. *Fracture and Fatigue Control in Structures: Application of Fracture Mechanics,* Prentice-Hall, Englewood Cliffs, NJ.)

5.5.3 CRITICAL CRACK LENGTH

For catastrophic failure, the initial crack depth a must reach the critical depth a_{cr}. This "critical crack depth" can be determined from Equation 5.53 by considering the critical stress intensity factor, K_{IC}, and the maximum applied stress, σ_{max}, as

$$a_{cr} = \left(\frac{K_{IC}}{1.1 M_k \sigma_{max}}\right)^2 \frac{Q}{\pi} \tag{5.54}$$

5.5.4 LEAK-BEFORE-BREAK

Leak-before-break (LBB) is a term first proposed by Irwin et al. [5] This concept is used widely to estimate the material fracture toughness, K_{IC}, required for a surface flow to grow through the thickness, t, thus allowing the pressurized components and vessels to fail from leakage prior to a fracture occurring in service. LBB has been applied to missile casings, gas and oil pipelines, pressure vessels, nuclear piping, and so on. As indicated in Figure 5.29, the first mode of failure (leakage) assumes that a

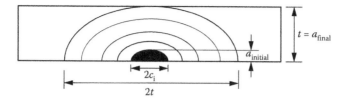

FIGURE 5.29 Leak-before-break criteria.

flaw twice the length of the wall thickness should be stable at a stress equal to the design stress [6]. That is, the critical crack size at the nominal design stress level of a material should be greater than the wall thickness.

Example 5.10

The design of the landing gear is one of the more essential aspects of aircraft design. The design and integration process requires many engineering disciplines, such as structures, weights, runway design, and economics, and has to be sophisticated when the aircraft size becomes larger. The landing gear design requirements create complexities in the development of a design methodology. These requirements include component maximum strength, minimum weight, high reliability, low cost, overall aircraft integration, airfield compatibility, and so on, and undoubtedly reflect the transdisciplinary nature of the task [7].

Figure 5.30 shows the main components of generic landing gear. Assume that the pressure inside the cylinder varies between 1 MPa and 20 MPa. If the inside diameter and the length of the cylinder are 0.3 m and 0.8 m, respectively, select

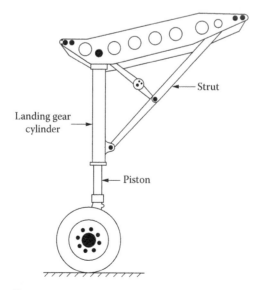

FIGURE 5.30 Landing gear.

TABLE 5.3

Materials for Landing Gear Cylinder

Material	S_y (MPa)	K_{IC} (MPa$\sqrt{m}$)	ρ (tons/m³)	E (GPa)	$ Cost/ton
Steel	1950	80	7.7	200	700
Aluminum alloys	500	24	2.7	70	800
Titanium alloys	1200	50	4.5	120	5400

the most suitable material from Table 5.3 using principles of fracture mechanics. Assume that the cylinder has a semielliptical surface crack and an initial surface flaw of depth 1 mm and an a/2c ratio of 0.25 that remains constant. Perform the analysis based on a maximum pressure of 20 MPa.

SOLUTION

First consider the analysis of steel. To determine the design stress, σ_d, that the pressure vessel can withstand, assume an allowable design stress of

$$\sigma_d = 0.5\, S_y$$

Assume that the stress intensity factor equation given by Equation 5.53 can be modified to determine the critical stress intensity factor as

$$K_{IC} = 1.1\, M_k \sigma_{max} \sqrt{\pi \frac{a}{Q}} \qquad (5.55)$$

Let $\sigma_{max} = \sigma_d$ and rearrange Equation 5.55 such that

$$\sigma_d = \frac{\sqrt{Q}.K_{IC}}{1.1 M_k \sqrt{\pi a}}$$

where

$$\sqrt{Q} = \sqrt{\Phi^2 - 0.212\left(\sigma_d/S_y\right)^2}$$

and

$$\Phi = \frac{3\pi}{8} + \frac{\pi}{8}\left(\frac{a}{c}\right)^2 = \frac{3\pi}{8} + \frac{\pi}{8}(0.5)^2 = 1.2763$$

$$\sqrt{Q} = \sqrt{(1.2763)^2 - 0.212(0.5)^2} = 1.2554$$

Since the design stress, σ_d, is a function of assumed values of M_k and ratio of σ_d/S_y, an iterative procedure should be used in the analysis. Assuming an error criterion, $\epsilon_{er} \leq 5\%$, the iteration process will continue until the converged value of design stress is reached. For the first iteration, assume that $M_k = 1$; hence,

$$\sigma_d = \frac{\sqrt{Q}K_{IC}}{1.1M_k\sqrt{\pi a}} = \frac{1.2554 \times 80}{1.1 \times 1\sqrt{\pi \times 0.001}} = 1629 \text{ MPa}$$

Using the thin-walled pressure vessel theory, we have

$$\sigma_d = \frac{pD}{2t} \text{ or } t = \frac{pD}{2\sigma}$$

$$\sigma_d = \frac{pD}{2t} \text{ or } t = \frac{pD}{2\sigma_d} = \frac{20 \times 0.3}{2 \times 1629} = 0.00184 \text{ m}$$

Then

$$\frac{a}{t} = \frac{0.001}{0.00184} = 0.543$$

Referring to Figure 5.28, we obtain $M_k = 1.086$. The new ratio of σ_d/S_y becomes

$$\frac{\sigma_d}{S_y} = \frac{1629}{1950} = 0.84$$

Using the new value of σ_d/S_y, recalculate $\sqrt{Q}$

$$\sqrt{Q} = \sqrt{(1.2763)^2 - 0.212(0.84)^2} = 1.216$$

Hence,

$$\sigma_d = \frac{1.216 \times 80}{1.1 \times 1.086\sqrt{\pi\, 0.001}} = 1453 \text{ MPa}$$

Compare the new calculated values with the old values to find the error as follows:

$$\text{Percent error, } \sigma_d = \left|\frac{1629 - 1453}{1629}\right| \times 100 = 10.8\%$$

$$\text{Percent error, } M_k = \left|\frac{1 - 1.086}{1}\right| \times 100 = 8.6\%$$

To gather initial guesses for the next iterations, bisect the range of σ_d/S_y and M_k. This will provide fast convergence.

$$\frac{\sigma_d}{S_y} = \frac{0.5 + 0.84}{2} = 0.67$$

$$M_k = \frac{1 + 1.086}{2} = 1.043$$

Following the same procedure, use $\sigma_d/S_y = 0.67$ and $M_k = 1.043$ as the initial values for the second iteration. After three iteration convergences are achieved and the results of analysis for steel are as follows:

$$t \quad = 1.9 \text{ mm}$$

$$\sigma_d = 1566 \text{ MPa}$$

$$\frac{\sigma_d}{S_y} = 0.79 \Rightarrow \text{ then safety factor, } n = \frac{S_y}{\sigma_d} = 1.25$$

$$M_k = 1.026$$

then the material cost of the landing gear cylinder out of steel is

$$\text{Cost} = m \times \$\text{cost/ton}$$

where

$$m = \rho \times V = \rho\left[\frac{\pi}{4}(D_o^2 - D_i^2) \times L\right] = 7.7 \times \left[\frac{\pi}{4}\left[(0.3 + 2 \times 0.0022)^2 - (0.3)^2\right] \times 0.8\right]$$

$$= 0.0128 \text{m}^3$$

Then the cost is

$$\text{Cost} = m \times \$\text{cost/ton} = 0.0128 \times 700 = \$8.96$$

Repeat the same calculations for the other two materials given in Table 5.3; the results are shown in Table 5.4. These results show that an increase in strength allows smaller thickness; thus, less material is used. An increase in toughness yields higher design stress, which provides better safety. Since aluminum alloy provides a design safety factor of almost unity, it should be eliminated from the list to be used as the candidate material. When the cost, weight, and safety factors are considered, steel is the only option. Although steel provides the highest design safety factor among the candidate materials, a safety factor of 1.25 is very low for this kind of design application. Of course, to have a reasonable safety factor, thickness can be increased.

TABLE 5.4
Calculation of Design Parameters

Material	S_y (MPa)	K_{IC} (MPa$\sqrt{m}$)	σ_d (MPa)	S_y/σ_d	t (mm)	ρ (tons/m³)	m (tons)	Total Cost ($)
Steel	1950	80	1566	1.25	1.9	7.7	0.0128	8.96
Aluminum alloys	500	24	465	1.08	6.5	2.7	0.0130	10.40
Titanium alloys	1200	50	985	1.22	3.3	4.5	0.0140	75.33

5.5.5 Fatigue Crack Propagation

In recent years, fracture-mechanics-based models for prediction of fatigue life have been used widely. As shown in Figure 5.31, fatigue life under cyclic load consists of three stages: the crack formation life, followed by a crack propagation period, and finally an unstable crack growth until failure occurs. Stage II represents the fatigue crack propagation and obeys the Paris power law. The power law between two given points is probably the simplest or most used form to predict fatigue life. The results of the crack growth tests of Paris were defined in terms of da/dN as a function of ΔK on a double log scale, which shows a linear relation between $\log(da/dN)$ and $\log(\Delta K)$ as shown in Figure 5.31. Several crack growth tests carried out by other researchers indicated the same pattern, which led to the well-known Paris power law equation:

$$\frac{da}{dN} = A(\Delta K_{\mathrm{I}})^{m} \tag{5.56}$$

where A and m are experimentally obtained constants for a particular material, environment, and loading condition, a is the crack length, N is the number of cycles, and ΔK_{I} is the stress intensity factor occurring at the crack tip.

The following steps are followed to predict fatigue life using fracture mechanics:

1. Assume an initial crack size, a_{i}
2. Determine the critical crack size, a_{cr}
3. Assume an incremental of crack growth, Δa

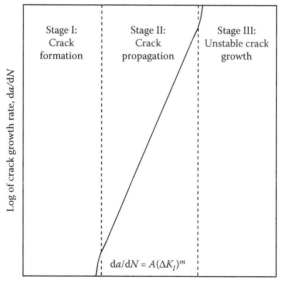

FIGURE 5.31   Fatigue-crack propagation stages.

4. Calculate ΔK_I using $\Delta\sigma$ and a_{av}, where

$$a_{av} = a_i + \frac{\Delta a}{2} \tag{5.57}$$

5. Calculate the number of cycles at a given stress level by using the Paris power law. Direct integration should continue until a reaches a_{cr}. If the thickness of the material is greater than a_{cr}, integration should stop when a reaches the material thickness.

Example 5.11

The tire bicycle pump shown in Figure 5.32a has an inside diameter of $D_i = 2.0$ in. and thickness of $t = 0.09$ in. For an initial flow size of 0.01 in., calculate whether the air chamber fails or leaks first, and estimate the number of life cycles if the stress varies in the chamber as shown in Figure 5.32b. The material has a yield strength of 3 kpsi and $K_{IC} = 2.5$ kpsi$\sqrt{in}$. Assume $A = 0.66 \times 10^{-8}$, $m = 2.25$, $M_k = 1$, and $a/2c$ remains constant.

SOLUTION

1. Assume that the air chamber has an elliptical surface flow (part through), as shown in Figure 5.33. Initial flow size is $a_i = 0.01$ in.
2. Calculate critical crack size, a_{cr}

$$a_{cr} = \left(\frac{K_{IC}}{1.1 M_k \sigma_{max}}\right)^2 \frac{Q}{\pi}$$

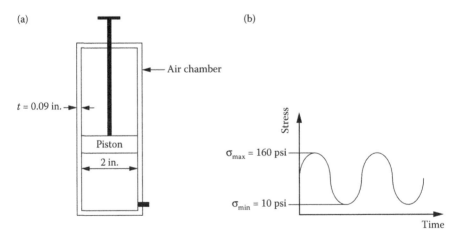

(a)   (b)

$t = 0.09$ in.

Air chamber

Piston

2 in.

$\sigma_{max} = 160$ psi

$\sigma_{min} = 10$ psi

Stress

Time

FIGURE 5.32 (a) Bicycle pump and (b) Stress variation in the chamber.

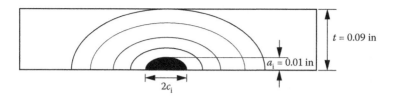

FIGURE 5.33 Semielliptical (part-through) surface crack.

where

$$\sqrt{Q} = \sqrt{\Phi^2 - 0.212\left(\sigma_d/S_y\right)^2}$$

and

$$\Phi = \frac{3\pi}{8} + \frac{\pi}{8}\left(\frac{a}{c}\right)^2 = \frac{3\pi}{8} + \frac{\pi}{8}(0.4)^2 = 1.2403$$

Substituting yields

$$\sqrt{Q} = \sqrt{1.2403^2 - 0.212\left(\frac{0.160}{3}\right)^2} \Rightarrow Q = 1.538$$

Then

$$a_{cr} = \left(\frac{K_{IC}}{1.1\, M_k \sigma_{max}}\right)^2 \frac{Q}{\pi} = \left(\frac{2.5}{1.1 \times 1.0 \times 0.160}\right)^2 \frac{1.538}{\pi} = 98.8\,\text{in.}$$

Since the critical crack depth, a_{cr} is greater than the thickness, the air chamber will leak before failure occurs.

3. Assume an increment of crack growth, Δa. Since $a_{cr} = 98.8$ in. $\geq t = 0.01$ in., integration should stop when a reaches the material thickness. For simplicity, assume two increments for crack growth, namely:

$$\Delta a = \frac{t - a_i}{2} = \frac{0.09 - 0.01}{2} = 0.04\,\text{in.}$$

4. Determine ΔK_I

$$a_{av} = a_i + \frac{\Delta a}{2} = 0.01 + \frac{0.04}{2} = 0.03\,\text{in.}$$

and

$$\Delta K_I = 1.1\Delta\sigma\sqrt{\pi\frac{a_{av}}{Q}} = 1.1 \times 0.150\sqrt{\pi\frac{0.03}{1.538}} = 0.04083\,\text{kpsi}\sqrt{\text{in.}}$$

5. Determine the number of cycles using

$$\frac{da}{dN} = A(\Delta K_i)^m \quad \text{where } A = 0.66 \times 10^{-8} \text{ and } m = 2.25.$$

To determine the incremental crack growth, the above equation can be modified to

$$\frac{\Delta a}{\Delta N} = A(\Delta K_i)^m$$

or

$$\Delta N = \frac{\Delta a}{A(\Delta K_i)^m} = \frac{\Delta a}{0.66 \times 10^{-8}(\Delta K_i)^{2.25}}$$

First iteration gives

$$\Delta N_1 = \frac{\Delta a}{0.66 \times 10^{-8}(\Delta K)^{2.25}} = \frac{0.04}{0.66 \times 10^{-8}(0.04083)^{2.25}} = 8.089 \times 10^9 \text{ cycles}$$

Second iteration gives

$$a_{av} = a_i + \frac{\Delta a}{2} = (0.01 + 0.04) + \frac{0.04}{2} = 0.07 \text{ in.}$$

Note that the initial crack size became 0.05" for the second iteration.

$$\Delta K_i = 1.1\Delta\sigma\sqrt{\pi \frac{a_{av}}{Q}} = 1.1 \times 0.150\sqrt{\pi \frac{0.07}{1.538}} = 0.06237 \text{ kpsi}\sqrt{\text{in.}}$$

and

$$\Delta N_2 = \frac{0.04}{0.66 \times 10^{-8}(0.06237)^{2.25}} = 3.118 \times 10^9 \text{ cycles}$$

Then the total cycle for failure is

$$\sum N = N_1 + N_2 = 8.089 \times 10^9 + 3.118 \times 10^9 = 1.1207 \times 10^{10} \text{ cycles}$$

Since the calculated total life $N = 1.1207 \times 10^{10}$ cycles is greater than $N = 10^6$ cycles, we assume that the air chamber will have infinite life.

5.6 VIBRATIONS IN DESIGN

5.6.1 KNOWLEDGE OF VIBRATIONS AND DESIGN ENGINEERS

A basic understanding of vibrating mechanical systems is important for design engineers. If structural vibration is not considered in structural design, catastrophic failures can result.

Vibration is important in most engineering design applications where the mechanical components are exposed to vibration during transport and service. The need to understand the effects of vibration on product reliability is vital where electronic/computer components are part of almost every product.

Most of us are familiar with vibration; a vibrating object oscillates back and forth. Some examples of vibrating objects in our daily lives are microwaves, which excite the food molecules at their natural frequency to heat up the food; guitar strings, which produce sound waves by a vibrating object; vehicles, which vibrate when driven on rough ground; and geological activity, which can cause enormous vibrations in the form of earthquakes.

Machine vibrations can take different forms. A machine component may vibrate over large or small distances, fast or slowly, and with or without detectable sound or heat. In general, machine vibrations are unintended and undesirable. Understanding the behavior of mechanical components using the vibration theory will lead the design engineer to better and more progressive designs of mechanical systems.

5.6.2 Terminology in the Field of Vibration

Mechanical System: A mechanical system is a group of interrelated or interdependent elements that have physical/force interactions.

Degree of Freedom: Degree of freedom is any of the minimum number of coordinates necessary to completely state the motion of a mechanical system.

Period: Period is the interval of time characterized by the incidence of a certain condition, event, or phenomenon.

Fundamental Frequency: Fundamental frequency (natural frequency) is the lowest frequency at which a system vibrates freely.

Resonance Frequency: A resonance frequency is a natural frequency of vibration determined by the physical parameters of the vibrating object.

Cycle: Cycle is the time between repeated events.

Amplitude: Amplitude is the *magnitude* of change in the oscillating variable with each *oscillation*.

Free Vibration: Free natural vibrations occur in a mechanical system when a body moves away from its position of rest. The internal forces lean to move the body back to its position of rest. In this case, the disturbing force $F(x, t)$ is equal to zero.

5.6.3 Natural Frequency of Spring–Mass System

A system possessing mass and spring is capable of free vibration, which takes place in the absence of external excitation. The natural frequency of the system is determined from the free vibration of the system. In Figure 5.34, the natural frequency of the system is the number of times mass m will oscillate (up and down) between its original position and its displaced position when there is no external excitation. In other words, if mass m is pulled downward and then released, the mass will oscillate at its natural frequency. Natural frequency is a function of the mass, m, and stiffness,

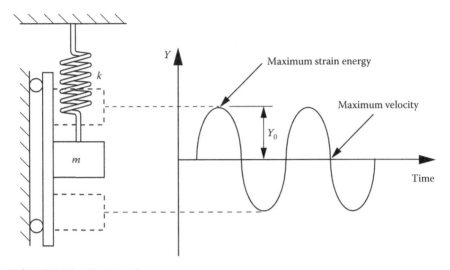

FIGURE 5.34 Simple spring–mass system.

k, of the system. Damping in fair amounts has little effect on the natural frequency of a system and, hence, can be neglected.

If a system shown in Figure 5.34 is subject to an external excitation that is close to its natural frequency, the spring can start to surge. This situation is very undesirable in design because not only can the life of the spring be reduced as high internal stresses result but also the operating characteristics of the spring can be seriously affected.

In engineering design, it is important to ensure that the natural frequency of vibration is not close to the excitation frequency. The Tacoma Narrows Bridge in Washington is a good example of this kind of situation. The bridge oscillated wildly because its natural frequency was very close to that caused by the wind. Eventually, this caused the bridge to fall into the river.

5.6.3.1 Derivation of Natural Frequency

Ignoring the mass of the spring and the damping of the system, the maximum kinetic energy, T_0, of the vibrating mass is given as

$$T_0 = \frac{1}{2} m V^2 \tag{5.58}$$

where velocity, V, is

$$V = \omega_n Y_0 \tag{5.59}$$

where ω_n is the natural frequency of the system. Substituting Equation 5.59 into Equation 5.58, we have the maximum kinetic energy as

$$T_0 = \frac{1}{2} m Y_0^2 \omega_n^2 \tag{5.60}$$

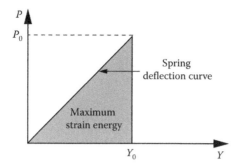

FIGURE 5.35   Strain energy of a spring.

Figure 5.35 shows the relationship between the spring deflection, Y, and the applied force, P. For a linear spring, the deflection is proportional to the applied force. The shaded area is the total maximum strain energy, U_0, and can be written as

$$U_0 = \frac{1}{2} P_0 Y_0 \qquad (5.61)$$

Force, P_0, as a result of a stretched spring is

$$P_0 = kY_0 \qquad (5.62)$$

After substituting Equation 5.62 into Equation 5.61, the maximum strain energy becomes

$$U_0 = \frac{1}{2} kY_0^2 \qquad (5.63)$$

When spring mass and damping are ignored, the kinetic energy will be equal to the strain energy at resonance

$$\frac{1}{2} mY_0^2 \omega_n^2 = \frac{1}{2} kY_0^2 \qquad (5.64)$$

Solving for natural frequency, ω_n,

$$\omega_n = \sqrt{\frac{k}{m}} \; \text{rad/s} \qquad (5.65)$$

Natural frequency in cycles per second (Hz) is

$$f = \frac{\omega_n}{2\pi} = \frac{1}{2\pi} \sqrt{\frac{k}{m}} \; \text{Hz} \qquad (5.66)$$

The spring rate, k, can also be written as

$$k = \frac{W}{\delta_{st}}$$ (5.67)

But mass, m, is given by

$$m = \frac{W}{g} \ (g = 32.2 \, \text{ft/s}^2)$$ (5.68)

Substituting Equations 5.67 and 5.68 into Equation 5.66 yields

$$f = \frac{1}{2\pi} \sqrt{\frac{g}{\delta_{st}}} \, \text{Hz}$$ (5.69)

The resonance frequencies of uniform beams and plates can also be determined by equating the kinetic energy to the strain energy at resonance. The results for various edge conditions are given in Figures A2.6 and A2.7 (see Appendix 2).

Example 5.12

Determine the natural frequency of the cantilevered beam with a steel tip mass as shown in Figure 5.36.

SOLUTION

Assume that the weight of the beam is small compared to the weight of the tip mass. Then the static deflection of the tip mass is

$$\delta_{st} = \frac{WL^3}{3EI}$$

where

$$I = \frac{\pi r^4}{2} = \frac{\pi \times 2^4}{2} = 25.12 \, \text{in.}^4$$

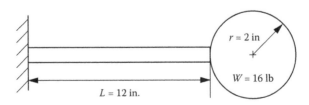

$r = 2$ in

$W = 16$ lb

$L = 12$ in.

FIGURE 5.36   Cantilever beam.

Then

$$\delta_{st} = \frac{WL^3}{3EI} = \frac{16 \times 12^3}{3 \times 30 \times 10^6 \times 25.12} = 0.01223 \times 10^{-3} \text{ in.}$$

$$f = \frac{1}{2\pi}\sqrt{\frac{g}{\delta_{st}}} = \frac{1}{2\pi}\sqrt{\frac{386}{0.01223 \times 10^{-3}}} = 895 \text{Hz}$$

5.6.4 DYNAMIC DISPLACEMENTS AND STRESSES

Many important approximations can be associated with the structural dynamic displacement developed during resonant conditions. Consequently, dynamic bending moment and stresses can be determined from dynamic displacements. From Figure 5.37, the vertical displacement can be written as

$$Y = Y_0 \sin \omega t \tag{5.70}$$

Taking the derivative of Equation 5.70 yields velocity

$$V = \dot{Y} = \omega Y_0 \cos \omega t \tag{5.71}$$

Taking the derivative of Equation 5.71 yields acceleration

$$a = \ddot{Y} = -\omega^2 Y_0 \sin \omega t \tag{5.72}$$

The negative sign in Equation 5.72 indicates that acceleration acts in the direction opposite to displacement. Maximum acceleration will occur when $\sin \omega t = 1$. Therefore, maximum acceleration becomes

$$a_{max} = \omega^2 Y_0 \tag{5.73}$$

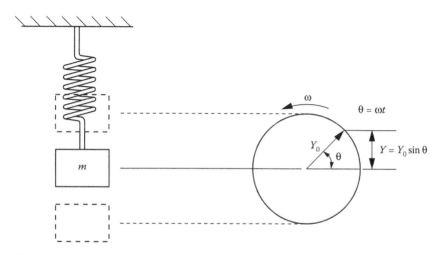

FIGURE 5.37 Simple harmonic motion of a mass–spring system.

where Y_0 is the maximum dynamic displacement. Defining acceleration in gravity unit, G

$$G = \frac{a_{max}}{g} = \frac{\omega^2 Y_0}{g}$$ (5.74)

But $\omega = 2\pi f$ and $g = 32.2$ ft/s^2 (386 in/s^2), then Equation 5.74 becomes

$$G = \frac{4\pi^2 f^2 Y_0}{g}$$ (5.75)

5.6.5 FORCED VIBRATION OF A SINGLE-DEGREE-OF-FREEDOM LINEAR SYSTEM

The response of a system to external excitation due to initial displacement, initial velocities, or both is an important subject in vibrations studies. When the external excitation is due to initial conditions (displacements or velocities) alone, the system response is called free vibration, as discussed previously. However, system excitation can also be in the form of forces that are applied for an extended period of time. Such vibration is called forced vibration and is commonly produced by the unbalance in rotating machines. Often, in engineering, this is known as harmonic excitation.

Harmonic excitation can be in the form of force or displacement. Now consider a single degree of freedom system with viscous damping, excited by harmonic force, $F = F_0 \cos \omega t$, as shown in Figure 5.38.

From the FBD (see Figure 5.38b), the equation of motion of the system is

$$m\ddot{y} + c\dot{y} + ky = F_0 \cos \omega t$$ (5.76)

Equation 5.76 consists of a complementary solution (transient solution) that dies out with time due to damping. On the other hand, the particular solution does not

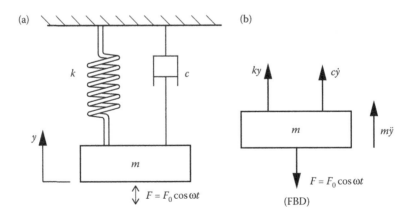

FIGURE 5.38 (a) Viscously damped system subject to harmonic excitation and (b) Harmonic excitation.

vanish in time and is known as the steady-state solution of the same frequency, ω, of the harmonic excitation. Then, the particular solution can be assumed in the form

$$y = y_0 \cos(\omega t - \phi) \tag{5.77}$$

where y_0 is the amplitude of the oscillation and ϕ is the phase of displacement with respect to the excitation. The amplitude of the oscillation, y_0, and the phase angle, ϕ, can be found by substituting Equation 5.77 into the Equation 5.76 as

$$y_0 = \frac{F_0}{\sqrt{\left(k - m\omega^2\right)^2 + \left(c\omega\right)^2}} \tag{5.78}$$

$$\phi = \tan^{-1} \frac{c\omega}{k - m\omega^2} \tag{5.79}$$

Equations 5.78 and 5.79 can be expressed in nondimensional form by dividing the numerator and denominator by k:

$$y_0 = \frac{F_0/k}{\sqrt{\left(1 - m\omega^2/k\right)^2 + \left(c\omega/k\right)^2}} \tag{5.80}$$

$$\tan\phi = \frac{c\omega/k}{1 - m\omega^2/k} \tag{5.81}$$

The above equation is further modified in terms of the following quantities:

$$\omega_n = \sqrt{\frac{k}{m}}$$

$$c_c = 2m\omega_n$$

$$\xi = \frac{c}{c_c}$$

$$R_\omega = \frac{\omega}{\omega_n}$$

where ω is the frequency of the excitation, ω_n is the natural frequency of the undamped oscillation, c_c is the critical damping, ξ is the damping factor, and R_ω is the frequency ratio. Then, the amplitude and phase angle become

$$y_0 = \frac{F_0/k}{\sqrt{\left(1 - R_\omega^2\right)^2 + \left(2\xi R_\omega\right)^2}} \tag{5.82}$$

and

$$\tan \phi = \frac{2\xi R_\omega}{1 - R_\omega^2}$$

$$T_r = \left[\frac{1 + \left(2R_\omega \xi\right)^2}{\left(1 - R_\omega^2\right)^2 + \left(2R_\omega \xi\right)^2} \right]^{1/2}$$ (5.83)

$$Q = \frac{1}{1 - R_\omega^2}$$

The ratio of the forced motion amplitude, y_0, to the static deflection, y_{st}, is often called a magnification factor. It signifies the amplitude of the forced vibration motion with respect to the magnification of the static deflection as a function of the frequency ratio R_ω. The magnification factor, A, becomes

$$A = \frac{y_0}{y_{st}} = \frac{1}{\sqrt{\left(1 - R_\omega^2\right)^2 + \left(2\xi R_\omega\right)^2}}$$ (5.84)

Figure 5.39 shows the dynamic amplification factor for several damping factors, which reveals that damping tends to diminish amplitudes and to shift the peaks to the left of the vertical at $R_\omega = 1$.

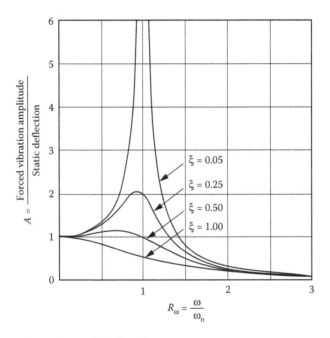

FIGURE 5.39   Dynamic amplification plots.

5.6.5.1 Vibration Isolation and Transmissibility

When a system is subjected to harmonic excitation, the support system also undergoes harmonic excitation. In rotating machines, as shown in Figure 5.40, unbalancing mass m with eccentric e, which is rotating with angular velocity, ω, creates harmonic excitation in the support system.

The purpose of vibration isolation is to control vibration so that its undesirable effects are kept within tolerable limits. In other words, the purpose of isolation is to reduce the vibration transmitted from the source to the support system and consequently to the foundation. In general, to isolate the transmitted vibration, springs (pneumatic, steel coil, rubber) are used. Natural frequency and damping are the basic properties of an isolator, which define the transmissibility of a support system. Transmissibility is the ratio of displacement of an isolated system to the input displacement or the ratio of force transmitted to a support system to the input disturbing force. As shown in Figure 5.41, transmissibility is used to describe the effectiveness of a vibration isolation system that varies with frequency.

In Figure 5.41, the region less than $\sqrt{2}$ around the resonance is the region of amplification. The force transmitted to the support system is less than the disturbing force only if the frequency ration is larger than $\sqrt{2}$. In other words, for smooth operation, the natural frequency of the support system must be considerably below the forcing frequency. At frequencies less than $\sqrt{2}$, no isolation takes place. Maximum transmissibility of an isolator occurs at resonance when the ratio of the forcing frequency to the natural frequency is equal to 1.

The transmissibility, T_r, of a dampened system can be expressed as

$$T_r = \left[\frac{1 + \left(2R_\omega \xi\right)^2}{\left(1 - R_\omega^2\right)^2 + \left(2R_\omega \xi\right)^2} \right]^{1/2} \tag{5.85}$$

The maximum transmissibility at resonance is generally referred to as the Q of the system and is approximately as follows: For lightly dampened systems, such as

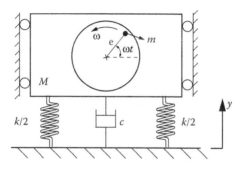

FIGURE 5.40 Rotating machine.

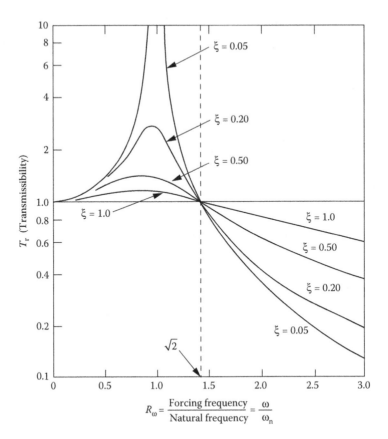

FIGURE 5.41 Family of transmissibility curves for a single-degree-of-freedom system.

those in Equation 5.85, damping factor ξ can be assumed to be zero. Then Q becomes

$$Q = \frac{1}{1 - R_\omega^2} \tag{5.86}$$

At resonance $R_\omega = 1$, Equation 5.85 can be reduced to

$$Q = \sqrt{\frac{1 + (2\xi)^2}{(2\xi)^2}} \tag{5.87}$$

For lightly dampened systems, the second term in the nominator of Equation 5.87 can be ignored

$$Q = \frac{1}{2\xi} \tag{5.88}$$

Equation 5.88 can also be written in terms of stiffness. Knowing the relationship of

$$\xi = \frac{c}{c_{cr}} = \frac{c}{2m\omega_n} = \frac{c}{2m\sqrt{k/m}} = \frac{c}{2\sqrt{km}} \tag{5.89}$$

and substituting into Equation 5.88 yields

$$Q = \frac{\sqrt{km}}{c} \tag{5.90}$$

Now, to determine the dynamic displacement, Y_{dyn}, Equation 5.75 is multiplied by Q

$$Y_{dyn} = \frac{386 \times G \times Q}{4\pi^2 \times f^2} \tag{5.91}$$

Example 5.13

As shown in Figure 5.42, a single-wheel small steam turbine used as a circulation pump has a total weight of 15 lb. Assume that the turbine rotor system is simply supported and a 15-lb concentrated force is acting at the center. If a sinusoidal vibration input created by unbalance force is 0.3 G, determine the life of the shaft carrying the turbine rotor. The S–N curve of the shaft material is shown in Figure 5.43. Use the following design data:

Distance between supports: $L = 8$ in.
Total weight: $W = 15$ lb
Shaft diameter: $D = 0.7$ in.
Elastic modules: $E = 30 \times 10^6$ psi
Material damping: $c = 0.6$ lb/in./s

SOLUTION

Figure 5.44 shows a single-degree-of-freedom representation of a rotor and shaft system. The mass represents the rotor, and the spring and dashpot represent the stiffness and the damping of the shaft.

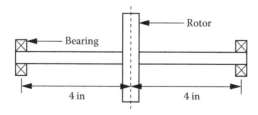

FIGURE 5.42 Steam turbine system.

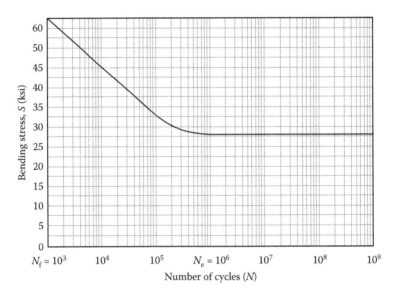

FIGURE 5.43 S–N curve.

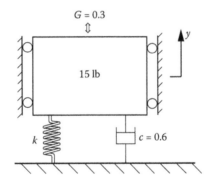

FIGURE 5.44 Single-degree-of-freedom representation.

From the strength of materials, the static deflection is

$$\delta_{st} = \frac{WL^3}{48EI}$$

where

$$I = \frac{\pi D^4}{64} = \frac{\pi(0.7)^4}{64} = 0.0118 \text{ in.}^4$$

Substituting yields

$$\delta_{st} = \frac{WL^3}{48EI} = \frac{15 \times 8^3}{48 \times 30 \times 10^6 \times 0.0118} = 4.52 \times 10^{-4} \text{ in.}$$

The natural frequency of the system is

$$f_n = \frac{1}{2\pi}\sqrt{\frac{g}{\delta_{st}}} = \frac{1}{2\pi}\sqrt{\frac{386}{4.52 \times 10^{-4}}} = 147 \text{ Hz}$$

For a lightly dampened system, transmissibility, Q, is

$$Q = \frac{1}{2\xi}$$

where damping ratio, ξ, is given by

$$\xi = \frac{c}{c_{cr}}$$

Critical damping, c_{cr}, is

$$c_{cr} = 2m\omega_n = 2m(2\pi f_n) = 2\left(\frac{15}{386}\right)(2\pi \times 147) = 72$$

Then

$$\xi = \frac{0.6}{72} = 0.0083$$

We now have the transmissibility, Q,

$$Q = \frac{1}{2 \times 0.0083} = 60$$

Dynamic displacement, Y_{dyn}

$$Y_{dyn} = \frac{386 \times G \times Q}{4\pi^2 \times f^2} = \frac{386 \times 0.3 \times 60}{4\pi^2 \times 147^2} = 8.153 \times 10^{-3} \text{ in.}$$

Dynamic displacement can also be calculated by modifying the static deflection to dynamic deflection as

$$Y_{dyn} = \frac{W_{dyn}L^3}{48EI}$$

where dynamic loading is

$$W_{dyn} = W \times G \times Q = 15 \times 0.3 \times 60 = 270 \text{ lb}$$

Substituting yields

$$Y_{dyn} = \frac{W_{dyn}L^3}{48EI} = \frac{270 \times 8^3}{48 \times 30 \times 10^6 \times 0.0118} = 8.136 \times 10^{-3} \text{ in.}$$

Note that the result is almost the same.

The maximum dynamic stress is determined by

$$\sigma_{dyn} = \frac{M_{dyn}c}{I}$$

The maximum dynamic bending moment, M_{dyn}, in the simply supported shaft is

$$M_{dyn} = \frac{W_{dyn}}{2} \times \frac{L}{2} = \frac{270}{2} \times \frac{8}{2} = 540 \text{ lb in.}$$

Then the maximum dynamic stress is

$$\sigma_{dyn} = \frac{540\left(\dfrac{0.7}{2}\right)}{0.0118} = 16,017 \text{ psi}$$

From the S–N curve shown in Figure 5.43, the total life is found to be an infinite number of cycles.

5.6.6 RESPONSE OF A SINGLE-DEGREE-OF-FREEDOM SYSTEM UNDER RANDOM EXCITATION

The vibration response spectrum is mainly used for random vibration inputs. Consider that a single-degree-of-freedom system, shown in Figure 5.45, is subjected to *white noise*, shown in Figure 5.46. Note that the white noise has a uniform density, S_0, for all frequencies.

The mean square of the response of a single-degree-of-freedom system shown in Figure 5.45 to white noise is given by [8]

$$\bar{G}^2 = \frac{1}{2\pi} \int_0^\infty \frac{S_0}{\left[1-\left(\omega/\omega_n\right)^2\right]^2 + 4\xi^2\left(\omega/\omega_n\right)^2} \, d\omega \qquad (5.92)$$

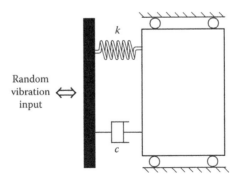

FIGURE 5.45 A single-degree-of-freedom with random input.

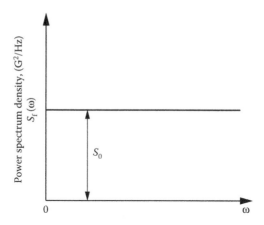

FIGURE 5.46 White noise vibration input.

Integration of Equation 5.92 by the method of residues yields

$$\bar{G}^2 = \frac{\omega_n S_0 Q}{4} \tag{5.93}$$

Substituting $\omega_n = 2\pi f_n$

$$\bar{G}^2 = \frac{\pi f_n S_0 Q}{2} \tag{5.94}$$

Then the root mean square (RMS) is the square root of Equation 5.94

$$\bar{G}_{RMS} = \sqrt{\frac{\pi}{2} f_n S_0 Q} \tag{5.95}$$

where S_0 is the power spectrum density input at the resonant frequency (in $\bar{G}^2$/Hz). The vibration response spectrum is usually represented in terms of acceleration response, $\bar{G}_{RMS}$, versus the natural frequency of the system. The resonant frequency (in Hz) is f_n, and Q is the transmissibility at the resonant frequency.

Gaussian distribution relates to the magnitude of the acceleration levels probable for random vibration. Assuming a Gaussian distribution, a simplified approach for evaluating random vibration fatigue life is the three-band technique:

1. Statistical probability of the response acceleration within $\pm 1\sigma$ values is 68.3%.
2. Statistical probability of the response acceleration within $\pm 2\sigma$ values is 27.1%.
3. Statistical probability of the response acceleration within $\pm 3\sigma$ values is 4.33%.

Example 5.14

In Example 5.13, if the turbine system subjected to the input created by the unbalance load is subjected to white noise vibration (see Figure 5.47) with a $S_0 = 0.01$ G²/Hz and a power spectrum density input from 20 to 300 Hz for the duration of 2 min, determine the remaining life of the shaft.

SOLUTION

As shown in Figure 5.47, the turbine assembly is subjected to a white noise whose power spectrum is almost flat in the neighborhood of the resonance frequency. Calculated design parameters from Example 5.13 are

$$I = 0.0118 \text{ in.}^4$$
$$f_n = 147 \text{ Hz}$$
$$Q = 60$$

RMS acceleration response of the shaft is

$$\bar{G}_{RMS} = \sqrt{\frac{\pi}{2} f_n S_0 Q} = \sqrt{\frac{\pi}{2} 147 \times 0.01 \times 60} = 11.77 \text{ RMS}$$

Dynamic loading is

$$W_{dyn} = W \times \bar{G}_{RMS} = 15 \times 11.77 = 176.6 \text{ lb RMS}$$

Dynamic bending moment is

$$M_{dyn} = W_{dyn} \times \frac{L}{2} = 176.6 \times \frac{8}{2} = 706.4 \text{ lb in RMS}$$

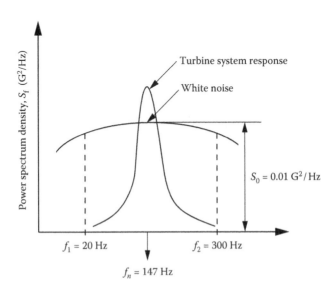

FIGURE 5.47   White noise vibration input and system response.

Dynamic bending stress is

$$\sigma_{dyn} = \frac{M_{dyn} \times D/2}{I} = \frac{706.4 \times 0.7/2}{0.0118} = 20,953 \text{ lb/in.}^2 \text{ RMS}$$

$\sigma_{dyn} = 20,953$ lb/in.^{2} RMS is at 1σ level input and will occur 68.3% of the time.
Dynamic stress at 2σ level is $\sigma_{dyn} = 2 \times 20,953 = 41,906$ lb/in.^{2} RMS
Dynamic stress at 3σ level is $\sigma_{dyn} = 3 \times 20,953 = 62,859$ lb/in.^{2} RMS

5.6.7 NUMBER OF ZERO CROSSING

"Zero crossing" is a point where the sign of a function changes from positive to negative. Thus, for fatigue life analysis, the number of stress reversals is determined from the number of positive zero crossing. For a single-degree-of-freedom system, the number of positive zero crossing is equal to its resonance frequency. Therefore, for this application the number of positive zero crossing is $n = f_n = 147$. Then the total number of stress cycles at each stress level can be calculated by

$$n = \text{(number of stress cycles, Hz)} \times \text{(duration time in seconds)} \times$$
$$\text{(\% occurance at each stress level)}$$

At 1σ level:

$$n_1 = (147) \times (2 \text{ min} \times 60 \text{ s/min}) \times (68.3\% \text{ occurance}) = 12,048 \text{ cycles}$$

From the S–N curve, fatigue life N_1 at $\sigma_{dyn} = 20,953$ lb/in^2 RMS stress level is infinite.

At 2σ level:

$$n_2 = (147) \times (2 \text{ min} \times 6 \text{ s/min}) \times (27.1\% \text{ occurance}) = 4780 \text{ cycles}$$

From the S–N curve, fatigue life N_2 at $\sigma_{dyn} = 2 \times 20,953 = 41,906$ lb/in^2 RMS stress level is 20,000 cycles.

At 3σ level:

$$n_3 = (147) \times (2 \text{ min} \times 60 \text{ s/min}) \times (4.33\% \text{ occurance}) = 764 \text{ cycles}$$

From the S–N curve, fatigue life N_3 at $\sigma_{dyn} = 3 \times 20,953 = 62,859$ lb/in^2 RMS stress level is 1500 cycles.
Using Miner's rule,

$$\left(\frac{n_1}{N_1} + \frac{n_2}{N_2} + \cdots + \frac{n_i}{N_i} \right) \times 100 = \left(\frac{12,048}{\infty} + \frac{4780}{20,000} + \frac{764}{1500} \right)$$
$$\times 100 = (0 + 0.239 + 0.509) \times 100 = 74.8\%$$

The remaining life of the shaft = $100 - 74.8 = 25.2\%$.

PROBLEMS

5.1.   A cantilevered bar shown in the following figure is made of steel and has a yield strength of $S_y = 80$ kpsi. Investigate whether a bar diameter of $D = 1.8$ in. is safe for a static safety factor of 2.

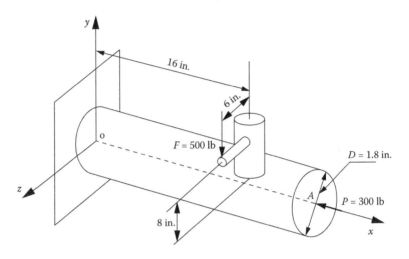

5.2.   A cantilevered bar shown in the following figure is made of steel and has a yield strength of 81 kpsi. The bar is subjected to $F = 400$ lb at point B; $T = 3000$ lb in. and $P = 600$ lb at point A. Investigate whether a bar diameter of $D = 1.5$ in. is safe for a static safety factor of 2. Ignore the direct shear effect.

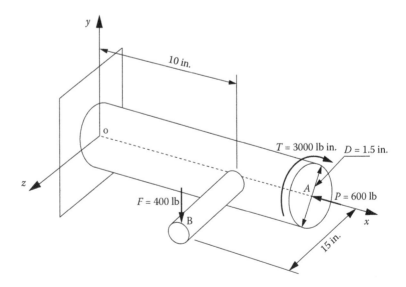

5.3. A cantilevered bar shown in the following figure is made of steel and has a yield strength of $S_y = 81$ kpsi. Investigate whether a bar diameter of $D = 1.5$ in. bar is safe for a static safety factor of 2.

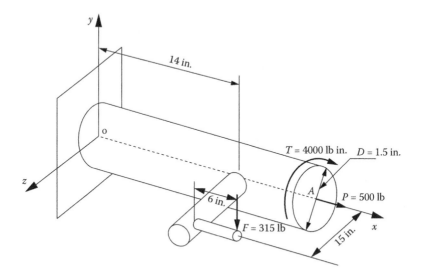

5.4. As shown in the following figure, there are external forces applied at point A along the centroidal axis of the bar and a constant force of F at point B. The bar is long and slender. Determine the force, F, that would cause the cantilevered bar to fail. Assume $S_y = 81$ kpsi.

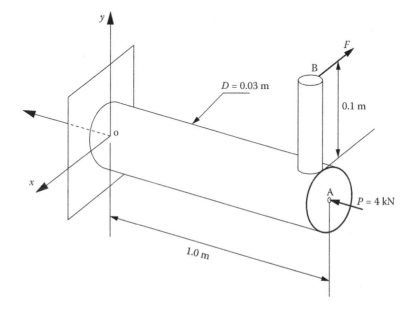

5.5.   If the ultimate strength of a steel bar is 100 kpsi, determine the uncorrected fatigue strength corresponding to a life of 80,000 cycles.

5.6.   If the stress, σ, in a machine part is 350 MPa, determine the expected fatigue life, N. Assume $S_{ut} = 600$ MPa.

5.7.   Using the following figure, develop an approximated S–N curve for a steel part that has ultimate strength of 800 MPa.

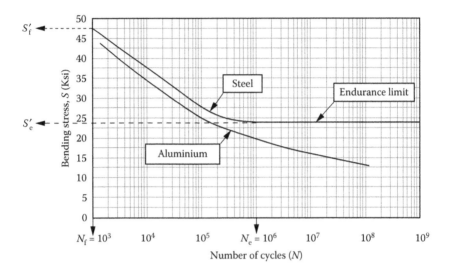

5.8.   Determine the corrected endurance limit of the steel part shown in the following figure for a reliability of 99.9. Assume that the shaft is machined and has $S_{ut} = 800$ MPa and $S_y = 400$ MPa and the material is ductile.

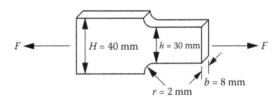

5.9.   A cantilevered shaft shown in the following figure is subjected to a load F varying from -5000 lb to $+5000$ lb. It is also subjected to a constant load of 20,000 lb along its centroidal axis. Assume that the shaft is manufactured as cold drawn steel with $S_y = 62$ kpsi and $S_{ut} = 70$

kpsi. Determine the acceptable diameter d for the shaft. Assume that the static safety factor is 3.

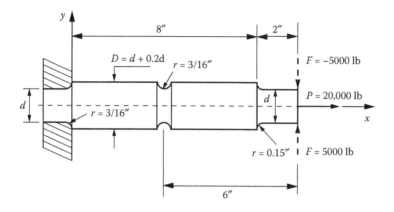

5.10. Resolve Example 5.7 if the fluctuating bending stress $\sigma_{max} = 100$ MPa and $\sigma_{min} = 0.0$ MPa. Assume that the material is brittle.

5.11. A fixed steel round bar shown in the following figure is subjected to a cyclic load of $P_1 = 1500$ lb for 500 cycles, then the magnitude of the cyclic load is reduced to $P_2 = 500$ lb for 10,000 cycles, and finally the cyclic load level is increased to $P_3 = 3000$ lb for 1,000 cycles. The steel bar is also subjected to a constant tension of $T = 4000$ lb. Assuming the material is brittle and the ultimate strength of the material is 90 kpsi, calculate the damage and remaining life of the steel bar.

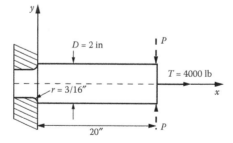

5.12. The following figure shows a round bar subjected to a fluctuating force P. The bar is made of steel (machined) and has an ultimate strength of 750 MPa and a yield strength of 470 MPa. Estimate the safety factor with respect to the eventual fatigue failure for 90% reliability. Assume that the material is ductile.

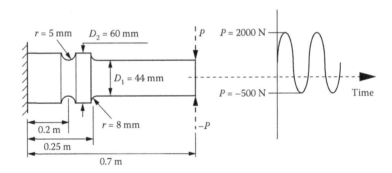

5.13. A drill bit with a diameter of 0.5 in. is used to make a hole in a composite wall, as shown in the following figure. Assume that the drill bit is rotating with a constant speed of 130 rpm throughout the composite wall and moving in the horizontal direction with a speed of 5 in./h and 7 in./h in concrete and steel plates, respectively. The drill bit is subjected to a constant compressive load of 1000 lb while drilling the composite wall, a bending moment of 500 lb in. while drilling through the concrete plate, and 400 lb in. while drilling through the steel plate. Calculate (a) the drill bit fatigue damage to make one hole in the composite wall; and (b) the number of holes that can be drilled before the drill bit fails.

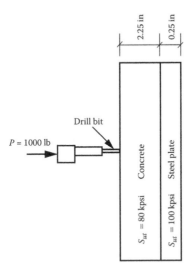

5.14. The following figure shows a shaft and repeated cyclic load through a camshaft at point C. The shaft is made of steel (machined) and has an ultimate strength of 800 MPa and a yield strength of 600 MPa. Estimate the factor of safety of the shaft for infinite life.

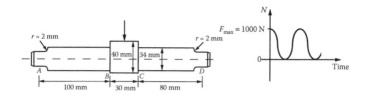

5.15. A torsion-bar suspension rod shown in the following figure is machined and has an ultimate strength of 800 MPa (116 kpsi). Assume that the torsion-bar is under pure, completely reversed torsion. If the safety factor is assumed to be 2, estimate the maximum value of torque that can be applied without causing fatigue failure.

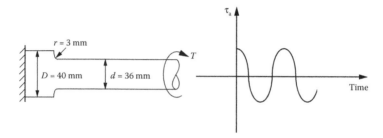

5.16. As shown in the following figure, a thin steel plate with finite width $2b = 100$ mm is subjected to a tensile stress. Assuming the critical stress intensity factor of the steel is $K_{IC} = 110 \, \text{MPa}\sqrt{m}$, calculate the tensile stress σ for failure. Assume that the thin plate has through-thickness crack $2a = 10$ mm.

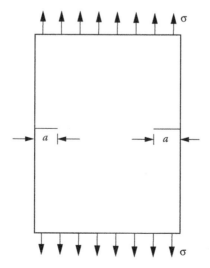

5.17. As shown in the following figure, one of the body center plates of a ship has a crack size of 0.83 in. Assume that the thickness of the steel plate is 1 in. Due to wave loads, the steel plate is subjected to a maximum bending stress of 50 kpsi and a minimum bending stress of 20 kpsi. If the critical stress intensity factor is 120 kpsi, calculate the life of the steel plate for catastrophic failure. Assume $A = 0.66 \times 10^{-8}$ and $m = 2.25$.

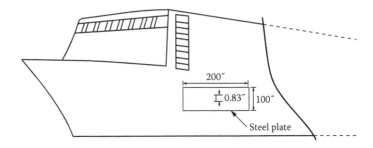

5.18. The following figure shows a 0.22 cal small bore stainless-steel barrel rifle with a diameter of 0.22 in. and a thickness of 0.05 in. As seen from the stress-time curve, when the rifle is fired the pulsating peak stress in the chamber reaches 70 kpsi. Determine the number of times one can fire the rifle before catastrophic failure occurs. The estimated initial flaw size a_i present in the thickness of the barrel is assumed to be 0.001. Material properties are: $K_{IC} = 65 \, \text{kpsi}\sqrt{\text{in.}}$, $S_y = 90 \, \text{kpsi}$.

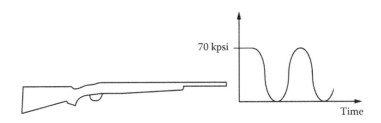

5.19. A cantilevered steel arm supports light assembly as shown in the following figure. The total weight of 10 lb of the light assembly is subjected to white noise as shown in the following figure. The power spectrum density input due to wind-induced vibration is $S_0 = 0.3 \, \text{G}^2/$ Hz from 5 to 40 Hz. If the random vibration duration is 10 min, determine the remaining life of the system. Assume that the steel bar has $S_{ut} = 80$ kpsi and damping of $c = 0.35$ lb/in/s.

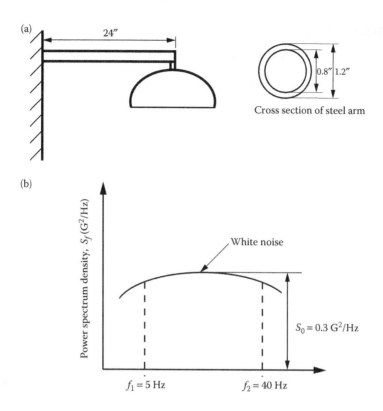

Cross section of steel arm

REFERENCES

1. Marin, J. 1962. *Mechanical Behavior of Materials*, Prentice-Hall, Englewood Cliffs, NJ, p. 224.
2. Shigley, J. E. and Mitchell, L. D. 1983. *Mechanical Engineering Design*, McGraw-Hill, New York.
3. Irwin, G. R. 1957. Analysis of stresses and strains near the end of a crack traversing a plate, *Transactions, ASME, Journal of Applied Mechanics*, 24, 361–364.
4. Irwin, G. R. 1962. The crack extension force for a part through crack in a plate, *Transactions, ASME, Journal of Applied Mechanics*, 29(4), 651–654.
5. Irwin, G. R. 1963. *Materials for Missiles and Spacecraft*, McGraw-Hill, New York.
6. Rolfe, S. T. and Barsom, J. M. 1977. *Fracture and Fatigue Control in Structures: Application of Fracture Mechanics*, Prentice-Hall, Englewood Cliffs, NJ.
7. Chai, S. T. and Mason, W. H. 1996. Landing gear integration in aircraft conceptual design, *MAD Center Report* (MAD 96-09-01), September.
8. Crandall, S. H. 1958. *Random Vibration*, John Wiley & Sons, Inc., New York.
9. Kobayashi, A.S., Zii, M., and Hall, L.R. 1965. Magnification factor Mk. *International Journal of Fracture Mechanics*, 1, 81–95.

BIBLIOGRAPHY

Ertas, A. and Jones, J. *The Engineering Design Process*, John Wiley & Sons. New York, 1996.

Juvinall, R. C. and Marshek, K. M. *Fundamentals of Machine Components Design*, John Wiley & Sons, New York, 2000.

Shigley, J. E. and Mitchell, L. D. *Mechanical Engineering Design*, McGraw-Hill, New York, 1983.

Steinberg, D. S. *Vibration Analysis for Electronic Equipment*, John Wiley & Sons. New York, 1988.

6 Design Analysis and Applications

6.1 INTRODUCTION

In this chapter, two tragic accidents that occurred in 2010 will be discussed, the first accident is the 2010 San Bruno pipeline explosion, which occurred on September 9, 2010, in San Bruno, California. This event killed at least five workers and caused more than 40 homes to burn to the ground. The second accident to be discussed is the *Deepwater Horizon* oil spill. This occurred on April 20, 2010 in the Gulf of Mexico and continued for three months and was the largest accidental marine oil spill in the history of the petroleum industry. The explosion killed 11 platform workers and injured many others. The spill has caused widespread damage to marine and wildlife habitats as well as the Gulf's fishing and tourism industries. The *Prevention through Design* concept and some related concerns will be discussed. The use of calculus knowledge in design analysis for practical engineering problems will also be discussed.

6.2 DESIGN ANALYSIS AND APPLICATION I

6.2.1 PIPELINE SYSTEM DESIGN

Extreme piping vibration can cause serious problems. Loose connections can cause leaking, and eventually pipes can be knocked off from their supports and in extreme cases, a pipe fatigue failure can occur. There are many causes of vibration in piping systems. Among them are rotating or reciprocating equipment with unbalanced force, pressure pulsations, flashing flow, and high fluid kinetic energy.

Pipe vibration is usually classified into two types: steady-state vibration and non-linear-transient vibration. Steady-state vibration is forced and cyclic and occurs over a long period of time. In many cases, steady-state vibration can cause a catastrophic failure in the pipe due to a large number of high stress cycles.

In general, nonlinear transient vibrations occur for a relatively short period of time and die quickly. A general transient piping vibration is a water hammer that may be caused by quick pump starts or stops or by fast valve closing or opening. During nonlinear transient vibrations of a piping system, pressure surges are induced and propagated at the speed of sound through the system. This transient response of fluid-filled pipes, usually referred to as a water hammer, involves large transient pressure pulses that can cause the pipe and its components to fail.

The determination of the natural frequencies of a pipeline system and the correct analysis of their behavior in the resonance case are important and necessary for the prevention and control of hazards and ensure the safety operation of the system.

The Office of Pipeline Safety (OPS) acknowledged that the majority of pipeline incidents are caused by "damage by outside force." Property damages alone for over 300,000 miles of transmission pipelines in the United States can cost operators millions of dollars every year [1]. A reliable technology and analysis process to test the pipelines is crucial to ensure the integrity of the pipeline system. For example, pipeline damage causing a crude oil leak would have a catastrophic effect on the environment. Pipeline damage leading to a gas leak would have a catastrophic effect on human life as well as incur millions of dollars property damage. Recently, a major gas pipeline explosion on September 9, 2010, in California left several dead and many homes burned to the ground in the Crestmoor residential neighborhood of San Bruno.

Responsible authorities and service providers must work actively to ensure the safety and reliability of the nation's pipeline transportation infrastructure. To minimize potential hazards and risk during the pipeline system design, preventative pipeline maintenance guidelines and pipeline system control, such as pipeline parameter monitoring, pipeline leak detection, fire and gas detection, and protection and emergency shutdown of the system, must be considered and documented. The functional requirements must be developed to define the required design life and design conditions. Anticipated normal, extreme, and shut-in operating settings with their potential ranges in flow rates, pressures, temperatures, fluid compositions, and fluid qualities should be identified and considered during the early stage of design development of the pipeline system.

6.2.2 RELATED DESIGN PROBLEM*

A simply supported steel pipeline has an inline pump at the center of the pipeline as shown in Figure 6.1. Due to the pump's dead weight and its operation at certain

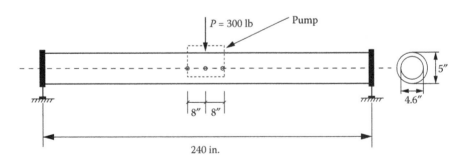

FIGURE 6.1   Pipeline system.

* Adapted from K. T. Truong, *Evaluating Dynamic Stresses of a Pipeline*, The Ultragen Group Ltd, Longueuil, Quebec, J4G 1G3.

rpm's, the steel pipeline is subjected to a harmonic force of 1335 N at its mid-span with an excitation frequency of 5 Hz.

Let us investigate whether the pipeline has infinite life of operation.

The undamped equation of motion of a piping system subjected to a harmonic forcing function with amplitude P_0 is

$$m\ddot{y} + ky = P_0 \sin \omega t \tag{6.1}$$

The assumed solution of Equation 6.1 is

$$y = y_0 \sin \omega t \tag{6.2}$$

Substituting Equation 6.2 into Equation 6.1, we obtain

$$-m\omega^2 y_0 \sin \omega t + ky_0 \sin \omega t = P_0 \sin \omega t \tag{6.3}$$

Dividing Equation 6.3 by $\sin \omega t$ yields

$$y_0 (k - m\omega^2) = P_0 \tag{6.4}$$

or

$$y_0 = \frac{P_0}{k - m\omega^2} \tag{6.5}$$

Dividing Equation 6.5 by k yields

$$y_0 = \frac{P_0 / k}{1 - \omega^2 / \omega_n^2} = \delta_{st} \frac{1}{1 - R_\omega^2} \tag{6.6}$$

where

$$R_\omega = \frac{\omega}{\omega_n}$$

and

$$\delta_{st} \frac{P_0}{k} \quad \text{(static deflection)}$$

Then the solution of Equation 6.1 can be obtained by substituting Equation 6.6 into Equation 6.2

$$y_0 = \delta_{st} \frac{1}{1 - R_\omega^2} \sin \omega t \tag{6.7}$$

Equation 6.7 provides the maximum dynamic displacement when $\sin \omega t = 1$, namely

$$\delta_{dyn} = \delta_{st} Q \tag{6.8}$$

where

$$Q = \frac{1}{1 - R_\omega^2} \qquad (6.9)$$

Assume that the piping system is a simply supported uniform beam with a concentrated force at the center. Then the natural frequency becomes (see Appendix 2, Figure A2.6)

$$f_n = \frac{1.57}{L^2} \sqrt{\frac{EIg}{w}} \qquad (6.10)$$

where
   Mass density, $\rho = 76.5$ kN/m³
   Total length, $L = 6.0$ m
   Young's modulus, $E = 207$ GPa
   Moment of inertia of the pipe cross section is

$$I = \frac{\pi \left(D_o^4 - D_i^4 \right)}{64} = \frac{\pi \left[\left(13 \times 10^{-2} \right)^4 - \left(12 \times 10^{-2} \right)^4 \right]}{64} = 383.9 \times 10^{-8} \text{ m}^4$$

Weight of the pipe is

$$W = \frac{\pi \left(D_o^2 - D_i^2 \right)}{4} \times L \times \rho = \frac{\pi \left[\left(13 \times 10^{-2} \right)^2 - \left(12 \times 10^{-2} \right)^2 \right]}{4} \times 6.0 \times (76.5 \times 10^3)$$

$$= 900.8 \text{ N}$$

Weight per unit length is

$$w = \frac{900.8}{6.0} = 150.1 \text{ N/m}$$

Then the natural frequency of the pipe is

$$f_n = \frac{1.57}{L^2} \sqrt{\frac{EIg}{w}} = \frac{1.57}{6.0^2} \sqrt{\frac{207 \times 10^9 \times 383.9 \times 10^{-8} \times 9.81}{150.1}} = 9.94 \text{ Hz}$$

Static deflection at the center of the pipe is

$$\delta_{st} = \frac{PL^3}{48EI} \qquad (6.11)$$

When $x < L/2$ the static deflection equation becomes

$$\delta_{st} = \frac{Px}{48EI}\left(3L^2 - 4x^2\right)$$ (6.12)

Note that the beam deflection is symmetrical with respect to the applied force position.

6.2.2.1 Calculation of Beam Dynamic Stress

From solid mechanics, the bending moment equation is

$$M = EI\frac{d^2y}{dx^2}$$ (6.13)

And the stress due to bending is given by

$$\sigma = \frac{MD}{2I}$$ (6.14)

Curvature d^2y/dx^2 in Equation 6.13 can be determined by using the finite difference method as shown in Figure 6.2.

Using the central difference method,

$$\frac{d^2y}{dx^2} = \frac{\left(y_{i-1}\right)-\left(2y_i\right)+\left(y_{i+1}\right)}{2\Delta x^2}$$ (6.15)

where Δx is the grid spacing as shown in Figure 6.2. The beam deflection at the load application point is y_i, where $x = L/2$. y_{i+1} and y_{i-1} are the beam deflections at $x = L/2 \pm \Delta x$. Substituting Equations 6.11 and 6.12 into Equation 6.15, we have

$$\frac{d^2y}{dx^2} = \frac{2P}{48EI \times \Delta x^2}\left(3L^2x - 4x^3 - L^3\right)$$ (6.16)

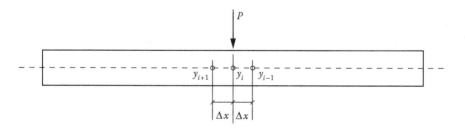

FIGURE 6.2 Finite difference model.

Substituting Equation 6.16 into Equation 6.13, we have

$$M = \frac{P}{24 \times \Delta x^2}\left(3L^2 x - 4x^3 - L^3\right) \tag{6.17}$$

Substituting Equation 6.17 into Equation 6.14, we have

$$\sigma = \frac{PD}{48 I \Delta x^2}\left(3L^2 x - 4x^3 - L^3\right) \tag{6.18}$$

For lightly damped systems, multiplying Equation 6.18 by Q gives the beam dynamic stress as

$$\sigma = \frac{PDQ}{48 I \Delta x^2}\left(3L^2 x - 4x^3 - L^3\right) \tag{6.19}$$

where

$$Q = \frac{1}{1 - R_\omega^2} = \frac{1}{1 - \left(5/9.94\right)^2} = 1.34$$

Assuming 30 elements for 6.0 m beam length, the grid spacing is

$$\Delta x = \frac{600}{30} = 20\,\text{cm}$$

Dynamic stress at $x = 280$ cm as shown in Figure 6.3 is

$$\sigma_{dyn} = \frac{1335 \times 0.13 \times 1.34}{48 \times 383.9 \times 10^{-8} \times 0.20^2}\left(3 \times 6^2 \times 2.8 - 4 \times 2.8^3 - 6^3\right) = 0.444 \times 10^8\ \text{N/m}^2$$

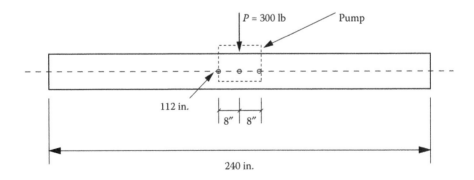

FIGURE 6.3   Maximum stress at $x = 280$ cm.

6.2.2.2 Calculation of Dynamic Stress

$$\frac{\delta_{dyn}}{\sigma_{dyn}} = \frac{(PL^3/48EI)\,Q}{PDQ/(48I\Delta x^2)(3L^2 x - 4x^3 - L^3)} \tag{6.20}$$

Rearranging

$$\delta_{dyn} = \sigma_{dyn}\left(\frac{L^3 \Delta x^2}{ED(3L^2 x - 4x^3 - L^3)}\right) \tag{6.21}$$

Substituting the value of σ_{dyn} into Equation 6.21 gives

$$\delta_{dyn} = 0.444 \times 10^8 \left(\frac{6^3 \times 0.20^2}{207 \times 10^9 \times 0.13(3 \times 6^2 \times 2.8 - 4 \times 2.8^3 - 6^3)}\right) = 102 \times 10^{-4}\,\text{m}$$

6.2.2.3 Fatigue Life Calculation

For $\sigma_{dyn} = 0.444 \times 10^8$ N/m^2 the fatigue life is

$$N = N_f\left(\frac{\sigma}{S_f'}\right)^{1/b} \tag{6.22}$$

where

$$b = \frac{\log\left(S_f'/S_e'\right)}{\log\left(N_f/N_e\right)} \tag{6.23}$$

Assuming that $S_{ut} = 689$ MPa

$$S_f' = 0.9 \times S_{ut} = 0.9 \times 689 = 620.1\,\text{MPa}$$
$$S_e' = 0.5 \times S_{ut} = 0.5 \times 689 = 344.5\,\text{MPa}$$
$$N_f = 10^3 \quad \text{and} \quad N_e = 10^6$$

Substituting into Equation 6.23, we have

$$b = \frac{\log\left(620.1/344.5\right)}{\log\left(10^3/10^6\right)} = -0.0851$$

Then

$$N = 10^3 \left(\frac{0.444 \times 10^8}{620.1 \times 10^6} \right)^{1/-0.0851} = 2.856^{16} \text{ cycles} \quad \text{(infinite life)}.$$

6.2.2.4 Questions as Closure
- What caused the San Bruno disaster to happen?
- What could the responsible people have done to prevent the hazard?
- What could the responsible people have done to control the hazard? The *Wall Street Journal* reported that it took almost two hours to manually cut off the natural gas that was fueling the firestorm.
- What are the lessons learned from this tragic accident?

6.3 DESIGN ANALYSIS AND APPLICATION II

6.3.1 OFFSHORE DRILLING

Owing to the scarcity of oil and gas in recent years, the exploitation of underwater resources has become necessary. This necessity created a number of challenging technical problems in the field of offshore technology. In particular, these problems have become of increasing concern for deepwater drilling systems under severe sea conditions. There are several different kinds of platforms for offshore drilling activities, from a fixed platform used for shallow water to drill ships able to operate in very deep waters.

An offshore drilling system is much more complex than shown in Figure 6.4 (not scaled). A marine riser is one of the most important components of the offshore drilling system. It is a structural connection between the blow-out preventer and the drilling platform and is used in drilling, production, and mining. Basically, a drilling riser is a cylindrical hollow steel tube that connects the drilling platform to the blow-out preventer located on the sea bottom. The drilling riser serves two main functions: first, it provides a pathway for the return of the drilling fluids (mud) to the drilling platform, and, second, it guides the drilling pipe from the surface to the well. For safety, environmental, and financial reasons, the riser must be structurally integral and reliable. Failure of the riser means costly down-time, considerable capital loss, and severe environmental pollution.

A blowout preventer is a bulky, specialized valve used to seal, control, and monitor oil and gas wells. Blowout preventers are crucial to the safety of the crew, drilling rig, and environment and to the monitoring and maintenance of well integrity; thus, blowout preventers are designed to be fail-safe devices. The failure of a blowout preventer means the loss of human life, expensive down-time, substantial capital loss, and severe environmental pollution. On April 20, 2010, a British Petroleum (BP) offshore oil rig exploded, killing 11 workers before the ultra-deepwater, semisubmersible deepwater horizon rig sank and spilled tens of thousands of barrels of crude oil into the Gulf of Mexico. The BP Deepwater Horizon blowout is the worst oil disaster and absolutely the worst environmental disaster in the history of the

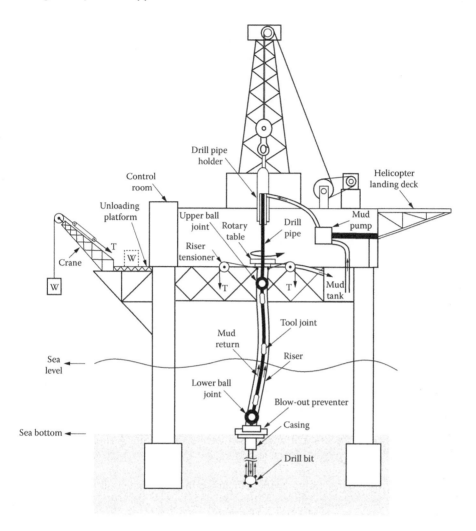

FIGURE 6.4 Fixed offshore drilling platform.

United States. Why and how did this happen? With a lot of speculation, Christopher Helman offered his thoughts on this issue (http://blogs.forbes.com/energysource/2010/04/30/bps-deepwater-disaster-what-happened-and-why/):

> We know with some certainty that workers were in the final stages of setting the final sections of pipe (production liner) in the hole and cementing it in place. The plan was to set cement plugs in the well, temporarily abandon it, and move the Deepwater Horizon off to a new drilling site within a couple days.
>
> Instead, it appears that a gas bubble got into the well bore, causing what's called a 'kick' as it traveled up the riser. Even a relatively small amount of gas can cause big problems because a gas bubble will expand massively as it moves from high pressure at the seafloor to lower pressure at the surface. The friction caused by the gas bubble

pushing up the pipe and displacing the drilling mud used to control the pressure could have created a static charge that ignited the gas into a fireball.

6.3.2 RELATED DESIGN PROBLEM*

As shown in Figure 6.4, a riser is one of the most important components of an off-shore drilling system. A riser is a long tensioned cylindrical hollow steel tube continuously subjected to wave and current forces. To minimize some of the forces due to the bending and to provide flexibility, the upper and lower ball joints are used in the upper and lower portion of the riser. It is apparent that the drill pipe undergoes a relatively sharp bend due to the deflection of the ball joints caused by hydrodynamic loads on the riser string. Bending stresses related to the bend at the ball joints can cause fatigue damage to the drill pipe inside the riser.

6.3.2.1 Derivation of Equations

Derive equations to calculate the bending moment and shear forces.

An element of the deformed drill pipe experiences a bending moment and shear forces as shown in Figure 6.5 while passing through the riser ball joint. To obtain the equations to calculate the bending moment and shear forces, consider the free-body diagram shown in Figure 6.5.

For equilibrium, the summation of the forces in the y direction should be equal to zero

$$\Sigma F_y = 0 + \leftarrow \tag{6.24}$$

$$-Q + (Q + dQ) = 0 \implies dQ = 0 \implies Q = \text{constant} \tag{6.25}$$

For equilibrium, the summation of the moments about point A should be equal to zero

$$\Sigma M_A = 0 \text{ (assuming counterclockwise moments are positive)} \tag{6.26}$$

$$M + dM - M + Q\, dx - T\, dy = 0 \tag{6.27}$$

Divide Equation 6.4 by dx

$$\frac{dM}{dx} - T\frac{dy}{dx} = -Q \tag{6.28}$$

Taking the derivative of Equation 6.28 with respect to x yields (Note that from Equation 6.25, the shear force Q is constant.)

$$\frac{d^2M}{dx^2} - T\frac{d^2y}{dx^2} = 0 \tag{6.29}$$

* Adapted from A. Ertas et al., 1989. *Transactions, ASME, Journal of Engineering for Industry*, 111(4), 369–374.

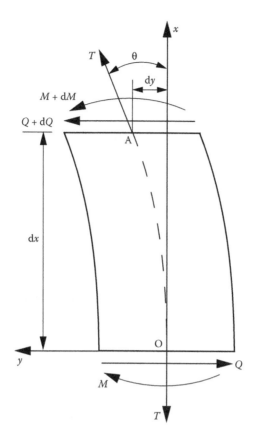

FIGURE 6.5 Free-body diagram of a drill pipe.

But the bending moment equation is

$$M = EI \frac{d^2 y}{dx^2} \tag{6.30}$$

Substituting Equation 6.30 into Equation 6.29 yields

$$EI \frac{d^4 y}{dx^4} - T \frac{d^2 y}{dx^2} = 0 \tag{6.31}$$

or

$$y'''' - \frac{T}{EI} y'' = 0 \tag{6.32}$$

Consider that tension, T, and EI are constant and let

$$k^2 = \frac{T}{EI} \tag{6.33}$$

Then Equation 6.32 becomes

$$y'''' - k^2 y'' = 0 \tag{6.34}$$

The solution of Equation 6.34 is

$$y = c_1 + c_2 x + c_3 \cosh(kx) + c_4 \sinh(kx) \tag{6.35}$$

Using Figure 6.6, boundary conditions can be written as

$$\text{(a)} \quad x = 0, \quad y = 0, \quad y' = 0, \quad \frac{d^2 y}{dx^2} = \frac{M}{EI}$$
$$\text{(b)} \quad x \to \infty, \quad M \to 0, \quad \frac{d^2 y}{dx^2} = 0, \quad y' = \frac{dy}{dx} \approx \theta \tag{6.36}$$

Applying boundary conditions for $x = 0$, $y = 0$, Equation 6.1 gives

$$c_1 + c_3 = 0 \implies \boxed{c_3 = -c_1} \tag{6.37}$$

Substituting Equation 6.37 into Equation 6.35 yields

$$y = c_1 + c_2 x - c_1 \cosh(kx) + c_4 \sinh(kx)$$

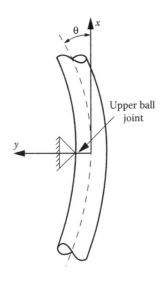

FIGURE 6.6   Upper ball joint model.

or

$$y = c_1[1 - \cosh(kx)] + c_2 x + c_4 \sinh(kx) \tag{6.38}$$

The derivative of Equation 6.38 with respect to x gives the slope

$$y' = \frac{dy}{dx} = -c_1 k \sinh(kx) + c_2 + c_4 k \cosh(kx) \tag{6.39}$$

Applying the boundary conditions of $x = 0$, $y' = 0$ yields

$$c_2 + c_4 k = 0 \implies \boxed{c_4 = -\frac{c_2}{k}} \tag{6.40}$$

Substituting Equation 6.40 into Equation 6.38, we have

$$y = c_1 \left[1 - \cosh(kx) \right] + \frac{c_2}{k} \left[kx - \sinh(kx) \right] \tag{6.41}$$

The second derivative of Equation 6.41 with respect to x gives

$$y'' = \frac{d^2 y}{dx^2} = k \left[-c_1 k \cosh(kx) - c_2 \sinh(kx) \right] \tag{6.42}$$

Considering the bending moment equation

$$M = EI \frac{d^2 y}{dx^2}$$

and applying boundary conditions at $x = \infty$, bending moment, $M = 0$, which yields $d^2 y/dx^2 = 0$, then

$$k[-c_1 k \cosh(kx) - c_2 \sinh(kx)] = 0 \tag{6.43}$$

Since k cannot be zero, terms in the brackets should be equal to zero, then

$$- c_1 k \cosh(kx) - c_2 \sinh(kx) = 0 \tag{6.44}$$

or

$$- c_1 k = c_2 \tanh(kx) \tag{6.45}$$

As seen from Figure 6.7, $\tanh(kx)$ is a function when $x = \infty$, $y = \tanh(kx) \approx 1$.

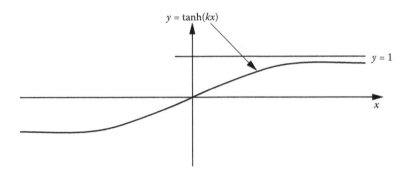

FIGURE 6.7   The tanh(kx) function.

Then from Equation 6.45, we have

$$c_1 = -\frac{c_2}{k}$$

(6.46)

Substituting Equation 6.46 into Equation 6.41

$$y = -\frac{c_2}{k}\left[1 - \cosh(kx)\right] + \frac{c_2}{k}\left[kx - \sinh(kx)\right]$$

(6.47)

Taking derivative of Equation 6.47

$$y' = \frac{dy}{dx} = c_2 \underbrace{\left\{\left[\sinh(kx) - \cosh(kx)\right] + 1\right\}}_{-e^{-kx}}$$

(6.48)

or

$$y' = c_2\left(-\frac{1}{1/e^{kx}} + 1\right)$$

(6.49)

Applying the boundary condition of $y' = 0$ when $x = \infty$, we obtain

$$\theta = c_2$$

(6.50)

Substituting Equation 6.50 into Equation 6.47 gives the deflection equation

$$y(x) = -\frac{\theta}{k}\left\{\left[1 - \cosh(kx)\right] - \left[kx - \sinh(kx)\right]\right\}$$

(6.51)

Derivative of Equation 6.51 gives the slope equation

$$y'(x) = \frac{dy}{dx} = c_2 \underbrace{\left\{ \left[\sinh(kx) + (1 - \cosh(kx)) \right] \right\}}_{-e^{-kx}}$$ (6.52)

Taking the derivative of Equation 6.52

$$y'' = \frac{d^2 y}{dx^2} = \theta k \left\{ \left[\cosh(kx) \right] - \left[\sinh(kx) \right] \right\}$$ (6.53)

And substituting Equation 6.53 into the moment equation of

$$M = EI \frac{d^2 y}{dx^2} \quad \text{gives}$$ (6.54)

$$M(x) = \frac{T\theta}{k} \left\{ \left[\cosh(kx) \right] - \left[\sinh(kx) \right] \right\}$$ (6.55)

where θ is the riser upper ball joint angle and

$$k = \sqrt{\frac{T}{EI}}$$ (6.56)

6.3.2.2   Operation Chart Design

By using fracture mechanics, construct an operation chart (variation of fatigue damage with respect to ball joint angle) when the drill pipe is subjected to a constant axial tension load of 60,000 lbs to prevent failure by buckling.

Design Data	
Module of elasticity of drill pipe material, E	30×10^6 psi
Critical stress intensity factor, K_{IC}	50 kpsi$\sqrt{\text{in.}}$
Yield strength of drill pipe material, σ_{ys}	80 kpsi
Crack growth (remains constant), a/c	0.5
Penetration speed of drill pipe, V	80 in./h
Outside diameter of drill pipe, D_o	5.00 in.
Inside diameter of drill pipe, D_i	4.276 in.
Drill pipe thickness, t	0.362 in.
Initial crack size, a_i	0.0312 in.
Fatigue life calculation constant, A	0.614×10^{-10}
Fatigue life calculation constant, m	3.16
Rotational speed of drill pipe, n	100 rpm
Tension load, T	60 kips

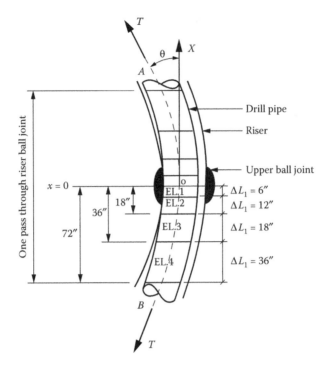

FIGURE 6.8 Schematic diagram of a drill pipe passing through the riser ball joint.

While the drilling operation is running, each element of the drill pipe shown in Figure 6.8 experiences alternate bending stress due to the rotation of the drill pipe. This bending stress is highest at the ball joint region and decays exponentially with distance from the ball joint, both below and above. As the drill pipe makes one pass through the ball joint, each element shown in Figure 6.8 will suffer cumulative damage from stress cycles of widely varying amplitudes.

6.3.2.3 Calculation of Constants Used in Computation

The moment of inertia of the drill pipe, I, is

$$I = \frac{\pi\left(D_o^4 - D_i^4\right)}{64} = \frac{\pi\left(5^4 - 4.276^4\right)}{64} = 14.2692 \text{ in.}^4$$

$T = 60{,}000$ lb and $E = 30 \times 10^6$ psi

$$k = \sqrt{\frac{T}{E \times I}} = \sqrt{\frac{60{,}000}{\left(30 \times 10^6\right)\left(14.27\right)}} = 0.0118 \text{ in.}^{-1}$$

The cross-sectional area of the drill pipe is

$$A = \frac{\pi\left(D_o^2 - D_i^2\right)}{4} = \frac{\pi\left(5^2 - 4.276^2\right)}{4} = 5.2746 \, \text{in.}^2$$

The mean stress, σ_m, due to tension applied on the drill pipe is

$$\sigma_m = \frac{T}{A} = \frac{60,000}{5.2746} = 11.3753 \, \text{kpsi}$$

6.3.2.4 Calculation of Cumulative Fatigue Damage

Determine the cumulative fatigue damage inflicted on the drill pipe during one pass through the ball joint. One pass is defined as the vertical movement of the drill pipe through 144 in. while being rotated. Assume that the drill pipe is subjected to a constant axial tension load of 60,000 lbs to prevent failure by buckling.

As shown in Figure 6.8, the drill pipe is divided into four elements during one pass through the riser ball joint. As seen from the figure, smaller elements are taken close to the higher-stress region. Since the fatigue damage will be considerably smaller at elements away from the ball joint, fatigue damage calculations will not be performed beyond 72 in. of the ball joint. The following are the steps to determine the fatigue damage:

Step 1: Calculate the alternating bending stress for $\theta = 1°$ and $T = 60,000$ lb. Using Equation 6.55, bending moment at $x = 0$ is

$$M(x = 0) = \frac{60,000}{0.01184} \times \left(1 \times \frac{\pi}{180}\right)\left\{\left[\cosh(0.01184)(0)\right] - \left[\sinh(0.01184)(0)\right]\right\}$$

$$= 88.453 \, \text{kips in.}$$

Bending moment at $x = 6$ is

$$M(x = 6) = \frac{60,000}{0.01184} \times \left(1 \times \frac{\pi}{180}\right)\left\{\left[\cosh(0.01184)(6)\right] - \left[\sinh(0.01184)(6)\right]\right\}$$

$$= 82.3878 \, \text{kips in.}$$

The average bending moment at element 1 is the average of bending moments at $x = 0$ and $x = 6$ in. That is,

$$M_{ave}(EL.1) = \frac{88.453 + 82.3878}{2} = 85.4204 \, \text{kips in.}$$

Then, the average alternating bending stress is

$$\sigma_{ave}(EL.1) = \frac{M_{ave}(EL.1) \times D_o}{2 \times I} = \frac{85.4204 \times 5}{2 \times 14.2692} = 14.9659 \, kpsi$$

Step 2: Determine the number of stress reversals, n_1, for the first element

$$n_1 = \frac{rpm \times 60 \times \Delta L_1}{V} = \frac{100 \times 60 \times (6-0)}{80} = 450 \, cycles$$

Step 3: Using the fracture mechanics approach, determine the number of cycles, N_1, for failure at the maximum stress level, σ_{max}, for the first element. Assume that the drill pipe has a semielliptical (part-through) surface crack and the estimated initial crack size, a_i, is 0.0312 in. Figure 6.9 shows the fluctuating stress time history acting on the drill pipe while passing through the ball joint.

From Figure 6.9, the maximum stress, σ_{max}, and stress range, σ_r, can be defined as

$$\sigma_{max} = \sigma_m + \sigma_{ave} = 11.3753 + 14.9659 = 26.3412 \, kpsi$$

and $\Delta\sigma$ is

$$\Delta\sigma = \sigma_r = 2\sigma_{ave} = \sigma_{max} - \sigma_{min}$$

Then

$$\Delta\sigma = 2 \times 14.9659 = 29.9318 \, kpsi$$

Step 4: Calculate the critical crack size, a_{cr}, that would cause catastrophic failure. Catastrophic failure occurs when the initial crack depth, a_i, reaches the critical crack size

$$a_{cr} = \left(\frac{K_{IC}}{1.1M_k\sigma_{max}}\right)^2 \frac{Q}{\pi}$$

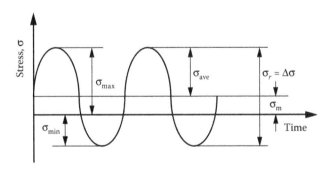

FIGURE 6.9 Fluctuating stress–time history of a drill pipe.

where

$$\sqrt{Q} = \sqrt{\Phi^2 - 0.212(\sigma_{max}/S_y)^2}$$

and

$$\Phi = \frac{3\pi}{8} + \frac{\pi}{8}\left(\frac{a}{c}\right)^2 = \frac{3\pi}{8} + \frac{\pi}{8}(0.5)^2 = 1.2763$$

Then the crack-shape parameter, Q, is

$$Q = (1.2763)^2 - 0.212(26.32/80)^2 = 1.6059$$

Assuming that $M_k = 1$, the critical crack size is

$$a_{cr} = \left(\frac{50}{1.1 \times 26.3412}\right)^2 \frac{1.6059}{\pi} = 1.5221\,\text{in.}$$

As shown in Figure 6.10, the critical crack size is greater than the wall thickness. Therefore, the drill pipe will leak before it fails. Using constants $A = 0.614 \times 10^{-10}$ and $m = 3.16$, the incremental life equation, ΔN, becomes

$$\Delta N = \frac{\Delta a}{A(\Delta K)^m} = \frac{\Delta a}{0.614 \times 10^{-10}(\Delta K_1)^{3.16}}$$

where

$$\Delta K_1 = 1.1\Delta\sigma\sqrt{\pi\frac{a_{av}}{Q}} = 1.95\Delta\sigma\sqrt{\frac{a_{av}}{Q}}$$

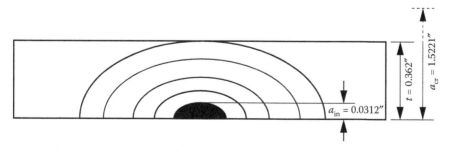

FIGURE 6.10 Critical crack size.

and

$$a_{av} = a_i + \frac{\Delta a}{2}$$

Substituting both ΔK_1 and a_{av} into the incremental life equation, ΔN, becomes

$$\Delta N = \frac{1.976 \times 10^9 \Delta a}{\left(\Delta\sigma \left(\frac{a_i + 0.5\Delta a}{Q} \right) \right)^{3.16}}$$

Assuming that incremental crack growth $\Delta a = 0.04$ in., the first iteration of the first element gives

$$\Delta N_1 (\text{EL.1}) = \frac{1.976 \times 10^9 \times 0.04}{\left(29.9318 \left(\frac{0.0312 + 0.5 \times 0.04}{1.6059} \right) \right)^{3.16}} = 395,960 \text{ cycles}$$

In case of the second iteration, using the initial crack size of $a_i = 0.0312 + 0.04 = 0.0712$ in., we obtain

$$\Delta N_2 (\text{EL.1}) = \frac{1.976 \times 10^9 \times 0.04}{\left(29.9318 \left(\frac{0.0712 + 0.5 \times 0.04}{1.6059} \right) \right)^{3.16}} = 159,040 \text{ cycles}$$

Since the calculated critical crack size is larger than the drill pipe thickness, the number of iterations will end when the drill pipe thickness, $t = 0.362''$, is reached. After eight iterations the total number of cycles for the first element, $\Sigma N(\text{EL.1})$, is

$$\sum N (\text{EL.1}) = 395,960 + 159,040 + 89,530 + 58,800 + 42,200 + 32,080$$
$$+ 25,400 + 20,730 + 20,550 = 844,280 \text{ cycles}$$

Note that for the last iteration, the incremental crack growth Δa is assumed to be 0.0508 in. to reach the exact thickness of 0.362 in. Then the damage for the first element is

$$\frac{n_1}{N_1} \times 100 = \frac{450}{834,700} \times 100 = 0.054\%$$

The above damage is multiplied by two because the same damage is inflicted on the drill pipe both above and below the riser ball joint. Then the total fatigue damage inflicted on the first element of the drill pipe during one pass through the riser upper ball joint is

$$2 \times \frac{n_1}{N_1} \times 100 = 2 \times \frac{450}{834,700} \times 100 = 0.108\%$$

The same procedure is followed for the remaining elements and a summary of the calculations is given in Table 6.1. It should be noted that since the stress on each element is changing, Q and a_{cr} should be recalculated for each element.

The total fatigue damage inflicted on the drill pipe during one pass through the riser upper ball joint when $\theta = 1°$ is

$$\text{Total damage} = 2 \times \left(\frac{n_1}{N_1} + \frac{n_2}{N_2} + \frac{n_3}{N_3} + \frac{n_4}{N_4} \right) \times 100 =$$

$$2 \times \left(\frac{450}{844,280} + \frac{900}{\infty} + \frac{1350}{\infty} + \frac{2700}{\infty} \right) \times 100 = 0.108\%$$

Fatigue damage calculations for $\theta = 1°$ is summarized in Table 6.1.

As seen from the above calculation, the damage is multiplied by two because the same damage is inflicted on the drill pipe both above and below the riser ball joint. Note that when the total life becomes larger than the infinite life cycle (10^6 cycles), the damage inflicted on the drill pipe is assumed to be zero. Figure 6.11 presents a summary of the damage calculations for $\theta = 1°, 2°, 3°, 4°, 5°$. Drilling operators must be alert when the riser upper ball joint angle exceeds $3°$. At any time after this point, drilling operation becomes dangerous and hazardous to human safety. Thus, drilling operation must be stopped. As seen from Figure 6.11, the fatigue damage inflicted on the drill pipe is exponentially growing with increasing upper ball joint angle, θ.

6.3.2.5 Offshore Platform Concrete Column Design

The offshore platform shown in Figure 6.4 is supported by four cylindrical hollow columns. Each column weighs 4.45×10^6 N. The outside and inside diameters of the columns are 4 and 3 m, respectively. The weight of the deck including equipment is

---

TABLE 6.1

Fatigue Damage Calculations

Upper Ball Joint Angle, θ	Element Number	σ_{av} (kpsi)	$\Delta\sigma$ (kpsi)	σ_{max} (kpsi)	n (cycles)	N (cycles)	Damage n/N (%)
$\theta = 1°$	1	14.97	29.93	26.34	450	844,280	0.108
	2	13.48	26.96	24.85	900	1.1781×10^6	0
	3	11.32	22.64	22.70	1350	2.0514×10^6	0
	4	8.36	16.73	19.74	2700	5.3622×10^6	0

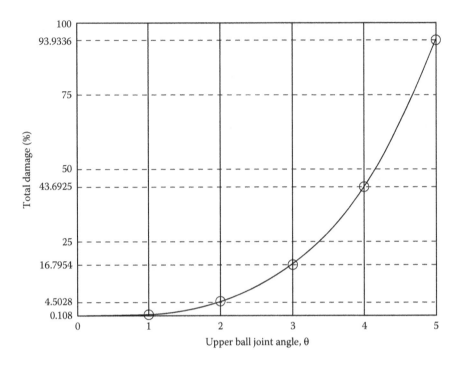

FIGURE 6.11   Drill pipe operating chart when $T = 60$ kpsi.

53.4×10^6 N. For each column, assume that the total force due to the wind, current, and wave is 1.335×10^6 N and is acting at 24 m above the sea bed. Assume that the compressive strength of the concrete, $S_{comp}^{con} = 35$ MPa.

1. Calculate the required cross-sectional area of the tendon, A^{ten}, for a given tendon stress of $\sigma_s^{ten} = 1100 \times 10^6$ N/m² to prevent the tensile stress effect in the concrete column.

The tendon cross section is given by

$$A^{ten} = \frac{A}{\sigma_s^{ten}}\left(\frac{Mc}{I} - \frac{F}{A}\right)$$

where $c = 2$ m, and

$$I = \frac{\pi\left(D0_o^4 - D_i^4\right)}{64} = \frac{\pi\left(4^4 - 2^4\right)}{64} = 11.78\,\text{m}^4$$

$$A = \frac{\pi\left(D_o^2 - D_i^2\right)}{4} = \frac{\pi\left(4^2 - 2^2\right)}{4} = 9.42\,\text{m}^2$$

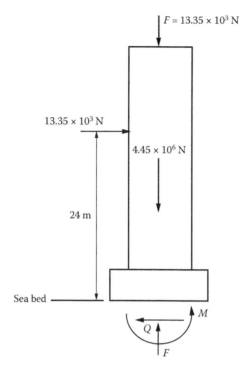

FIGURE 6.12 Free-body diagram.

The bending moment acting on the column is (See Figure 6.12)

$$M = 24 \times 1.335 \times 10^6 = 32.52 \times 10^6 \text{ m N}$$

The compressive force due to the weight of the column and the deck is

$$F = \frac{53.4 \times 10^6}{4} + 4.45 \times 10^6 = 17.8 \times 10^6 \text{ N}$$

and then the tendon cross-sectional area is

$$A^{\text{ten}} = \frac{A}{\sigma_s^{\text{ten}}} \left(\frac{Mc}{I} - \frac{F}{A} \right) = \frac{9.42}{1100 \times 10^6} \left(\frac{32.52 \times 10^6 \times 2}{11.78} - \frac{17.8 \times 10^6}{9.42} \right) = 0.0311 \text{m}^2$$

This cross-sectional area can be divided into equal numbers of tendons.

2. Calculate the maximum compressive stress, $\sigma_{\text{max}}^{\text{con}}$, in the concrete column.

$$\sigma_{\text{max}}^{\text{con}} = -\frac{Mc}{I} - \frac{F}{A} - \frac{\sigma_s^{\text{ten}} \times A^{\text{ten}}}{A} = \left(\frac{32.52 \times 10^6 \times 2}{11.78} - \frac{17.8 \times 10^6}{9.42} \right) - \frac{1100 \times 10^6 \times 0.0311}{9.42}$$

$$= -11.042 \times 10^6 \text{ N/m}^2$$

Assuming that $S_{comp}^{con} = 35\,\text{MPa}$, the maximum allowable compressive stress is

$$\sigma_{all}^{con} = 0.45 \times S_{comp}^{con} = 0.45 \times 35 = 15.75\,\text{MPa}$$

Then the safety factor is

$$n = \frac{15.75}{11.042} = 1.43$$

3. Calculate the maximum principal stress in the concrete column.

The maximum shear stress is given by

$$\tau_{max}^{con} = \frac{V\overline{Q}}{Ib}$$

where V is the maximum transverse shear force, I is the area moment of inertia about the neutral axis, b is the width of the beam, and $\overline{Q}$ is the first moment of the cross-sectional area on which shearing stress occurs. $\overline{Q}$ is given by

$$\overline{Q} = \int y\,dA = \overline{y}\overline{A}$$

where $\overline{A}$ is the shear area above the neutral axis as shown in Figure 6.13, $\overline{y}$ defines the location of the centroid of the shear area with respect to the neutral axis.

The location of the centroid of the shear area with respect to the neutral axis is

$$\overline{y} = \frac{\sum A\overline{y}}{\sum A} = \frac{A_1 y_1 - A_2 y_2}{A_1 - A_2} = \frac{\left((\pi/2)r_1^2\right)\left((4/3\pi)r_1\right) - \left((\pi/2)r_2^2\right)\left((4/3\pi)r_2\right)}{(\pi/2)r_1^2 - (\pi/2)r_2^2}$$

$$= \frac{2/3\left(r_1^3 - r_2^3\right)}{\pi/2\left(r_1^2 - r_2^2\right)}$$

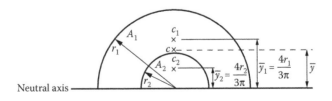

FIGURE 6.13   Shear area above the neutral axis.

Then $\bar{Q}$ is

$$\bar{Q} = A\bar{y} = \frac{\pi}{2}\left(r_1^2 - r_2^2\right) \times \frac{2/3\left(r_1^3 - r_2^3\right)}{\pi/2\left(r_1^2 - r_2^2\right)} = \frac{2}{3}\left(r_1^3 - r_2^3\right) = \frac{2}{3}\left(2^3 - 1^3\right) = 4.67\,\text{m}^2$$

Maximum shear stress is

$$\tau_{\text{max}}^{\text{con}} = \frac{V\bar{Q}}{Ib} = \frac{1.335 \times 10^6 \times 4.67}{11.78 \times (4-2)} = 0.26462 \times 10^6\,\text{N/m}^2$$

The normal stress along the bending axis is

$$\sigma = -\frac{F}{A} - \frac{\sigma_s^{\text{ten}} \times A^{\text{ten}}}{A} = -\frac{17.8 \times 10^6}{9.42} - \frac{1100 \times 10^6 \times 0.0311}{9.42} = -5.52123 \times 10^6\,\text{N/m}^2$$

Then the maximum principal stress becomes

$$\sigma_1 = \frac{\sigma}{2} + \sqrt{\left(\frac{\sigma}{2}\right)^2 + \tau^2} = -\frac{5.52123 \times 10^6}{2} + \sqrt{\left(\frac{5.52123 \times 10^6}{2}\right)^2 + (0.26462 \times 10^6)^2}$$

$$= 12.653 \times 10^3\,\text{N/m}^2$$

Comparing the critical fracture stress, $\sigma_{\text{cr-frac}}^{\text{con}}$, with the maximum principal stress

$$\sigma_{\text{cr-frac}}^{\text{con}} = 4\sqrt{S_{\text{comp}}^{\text{con}}} = 4\sqrt{35} = 23.664 \times 10^6\,\text{N/m}^2$$

The critical fracture stress value is much larger than the maximum principal stress.

6.3.2.6   Design Analysis of the Drill Pipe Holder Mechanism

The 1000 kg drill pipe is elevated by the drill pipe holder mechanism as shown in Figure 6.14. The drums constitute a single unit with 200 kg mass and rotating about the bearing at point O. If the radius of gyration of the drums about point O is 0.3 m

Design Parameters	
Weight of the drums	$W_{\text{dr}} = 200 \times 9.81 = 1962\,\text{N}$
Weight of the drill pipe	$W_{d-p} = 1000 \times 9.81 = 9810\,\text{N}$
Radius of gyration of the drum	$k = 0.5\,\text{m}$
Force applied by the motor	$F = 2.0\,\text{kN}$

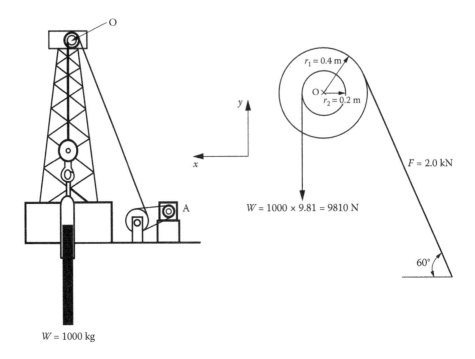

$W = 1000$ kg

FIGURE 6.14   Drill pipe holder mechanism.

and a constant force of 2.0 kN is maintained by the motor at A, determine the force components at the bearing O.

$$I_o = k^2m = (0.3)^2 \times 200 = 18 \text{ kg m}^2 \qquad (1)$$

Taking the moment about point O yields

$$\Sigma M_o = I_o\alpha \qquad (2)$$

$$-T \times r_1 + F \times r_2 = I_o\alpha \qquad (3)$$

or

$$-T \times 0.2 + 2.0 \times 10^3 \times 0.4 = 18\alpha \qquad (4)$$

From Figure 6.15, the summation of forces in the y direction yields

$$\Sigma F = ma_y \qquad (5)$$

$$T - 9810 = 1000a \qquad (6)$$

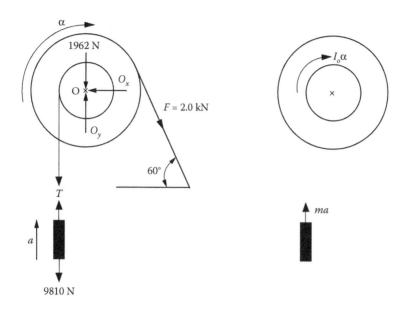

FIGURE 6.15 Free-body diagram.

but

$$a_t = r\alpha = 0.2\alpha \tag{7}$$

From Equations (6) and (7)

$$T = 9810 + 200\alpha \tag{8}$$

Substituting Equation (8) into Equation (4) and solving for α yields

$$-(9810 + 200\alpha) \times 0.2 + 2.0 \times 10^3 \times 0.4 = 18\alpha$$

$$\alpha = 52.8 \text{ rad/s}^2$$

Then

$$T = 9810 + 200 \times 52.8 = 20{,}370 \text{ N}$$

and the acceleration is

$$a = 0.2 \times 52.8 = 10.56 \text{ m/s}^2$$

The bearing reaction force components at point O are

$$\Sigma F_x = 0$$

$$O_x - 2.0 \times 10^3 \times \cos 60 = 0 \implies O_x = 1000 \text{ N}$$

$$\Sigma F_y = 0$$

$$O_y - 1962 - 2.0 \times 10^3 \times \sin 60 - 20{,}370 = 0 \implies O_y = 24{,}064 \text{ N}$$

6.3.2.7 Safety Concerns for Drilling Mud Pumps

An essential part of offshore drilling, drilling rig mud pumps, circulate the drilling fluids (mud) that are used to facilitate the drilling of oil and natural gas wells. Mud pumps are used to stabilize the pressure and to support the well during the drilling process. The drilling mud also helps in cooling and lubricating the drill bit. A mud pump circulates drilling fluid from the mud tanks, through the top drive, down the drill string, and through the bit. After the mud exits the drilling bit, it travels through the riser back to the surface carrying the cuttings made by the bit. To remove the cuttings, drilling mud passes through a shale-shaker and clean mud returns to the mud tank, and the clean drilling mud is ready for use again.

To prevent the air lock occurring for low pump inlet pressure, every drilling rig mud pump system includes a complete charging system. A charging system consists of a charging pump, the pump base, a butterfly valve, and the corresponding manifold. The charging pump is mounted on the suction manifold of the mud pump.

Consider now an accident on a fixed offshore drilling platform. A 30-year-old worker was injured on April 8, 1996 (http://www.larrycurtis.com/aop/louisiana-offshore-platform-accidents/). On the day of the accident, the worker was working as a member of a company's casing crew.

The worker who met with the accident stated that the responsible company did not provide a "charging pump" with the drilling rig mud pump system. The evidence established that the "charging pump" was not made available for the casing operation. The "charging pump" is a low-pressure pump relative to the rig mud pump. Another worker, a "tool-pusher," decided to connect the mud line to the rig mud pump that has very high-pressure discharge.

The worker who had the accident contended that the individuals involved in this casing operation had little, if any, experience using a rig mud pump. The worker who had the accident specifically argued that the driller on tour had virtually no experience in an operation like this. He also argued that notwithstanding the driller's lack of experience, this high-pressure rig mud pump was under the sole and exclusive control of the driller on tour; that is, the driller on tour was the only individual involved in this operation who could control the speed of the mud passing through the mud line.

As the work progressed, the plaintiff would signal the driller to turn the rig mud pump on and off. This would cause the mud flow to start and stop. At the very moment of the accident, the plaintiff had signaled the driller to activate the rig mud pump. The driller responded by turning the mud pump "on," but he caused the rig mud pump to be run "wide open." The force of the mud hitting the plaintiff caused him to be propelled more than 15 ft into the derrick. As a result of this accident, the worker who met with the accident suffered a severe crush injury to his right hand and injuries to his cervical and lumbar spine.

The worker who had the accident alleged that the defendant company was liable for his accident and the resulting injuries because of its negligence in failing to provide the proper pump to facilitate safe performance of the work and in failing to assign an experienced driller to the project. He additionally alleged that the defendant company was vicariously liable for the negligence of its employee driller in allowing the rig mud pump to be run wide open. His expert petroleum engineer testified that the manner in which the rig mud pump was lined up on the fill-up line at the time of the accident in question constituted a hazardous condition and gave rise to an unsafe workplace. This expert also maintained that a low-pressure charging centrifugal mud pump should have been lined up on the casing fill line.

If we accept the above scenario, this accident could have been prevented by using an experienced driller and also by designing protective connectors that connect the pump discharge to the mud line. In other words, a high-pressure pump discharge should have been designed in such a way that it should not fit in the low-pressure mud line.

6.3.2.8 Shaft Design

For safe operation, vibration analysis of a charging mud pump is extremely important to a designer. As shown in Figure 6.16, a centrifugal pump used for mud circulation has a total weight of 3100 N. Assume that the pump impeller system is simply supported and has a 3100 N concentrated force acting at the center. If the running speed of the pump is 1700 rpm, determine the safe shaft diameter, d.

The tentative design calls for the natural frequency of the pump to be twice the running speed (see Section 5.6.3). We know that the deflection formula for a simply supported beam is

$$\delta_{st} = \frac{WL^3}{48EI} = \frac{3100 \times (0.08 + 0.08)^3}{48 \times 207 \times 10^9 \left(\pi d^4 / 64 \right)} = \frac{0.026047}{d^4 \times 10^9}$$

The natural frequency is

$$f_n = \frac{1}{2\pi} \sqrt{\frac{g}{\delta_{st}}} = \frac{1}{2\pi} \sqrt{\frac{9.81}{0.026047/d^4 \times 10^9}} = 9.77 \times 10^4 \times d^2 \text{ Hz}$$

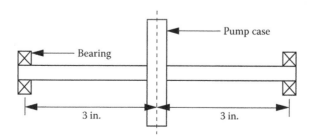

FIGURE 6.16 Centrifugal charging mud pump.

The pump speed in terms of cps is

$$\frac{1700}{60} = 28.33\,\text{cps}$$

For a safe operation, consider twice the running speed of the pump (see Chapter 5); then the safe diameter is

$$28.33 \times 2 = 9.77 \times 10^4 \times d^2 \implies d = 0.024 \text{ m}$$

6.3.2.9 Questions as Closure

- What caused the British Petroleum (BP) offshore oil rig to explode?
- What could the responsible people have done to prevent the hazard?
- What could the responsible people have done to control the hazard?
- What are the lessons learned from this tragic accident?

REFERENCES

1. http://www.ftpemea.com/pipeline_protection_case_studies.htm
2. K. T. Truong, *Evaluating Dynamic Stresses of a Pipeline*, The Ultragen Group Ltd, Longueuil, Quebec, J4G 1G3.
3. A. Ertas et al., 1989. The effect of tool joint stiffness on drill pipe fatigue in riser ball joints, *Transactions, ASME, Journal of Engineering for Industry* 111(4), 369–374.

Appendix 1: Tables

TABLE A1.1
Standard SI Prefixes

Factor	Prefix	Symbol
10^1	deca	da
10^2	hecto	h
10^3	kilo	k
10^6	mega	M
10^9	giga	G
10^{12}	tera	T
10^{-1}	deci	d
10^{-2}	centi	c
10^{-3}	milli	m
10^{-6}	micro	μ
10^{-9}	nano	n
10^{-12}	pico	p

TABLE A1.2
SI Base Units

Measurement	Abbreviations	Units
Length	l, s	meter (m)
Mass	M	kilogram (kg)
Time	t	seconds (s)
Electric current	I	ampère (A)
Temperature	T	kelvin (K)
Amount of substance	n	mol (mol)
Luminous intensity	l	candela (cd)

TABLE A1.3
SI Units in Terms of Base Units

Measurement	Abbreviations	Units
Area	Square meter	m^2
Volume	Cubic meter	m^3
Speed	Velocity meter per second	m/s
Acceleration	Meter per second squared	m/s^2
Wavenumber	Reciprocal meter	m^{-1}
Density, mass density	Kilogram per cubic meter	kg/m^3
Specific volume	Cubic meter per kilogram	m^3/kg
Current density	Ampere per square meter	A/m^2
Magnetic field strength	Ampere per meter	A/m
Concentration (of amount of substance)	Mole per cubic meter	mol/m^3
Luminance	Candela per square meter	cd/m^2

TABLE A1.4
SI Conversion of US Units to SI Units

Convert from	To	Multiply
atmosphere (standard)	Pa (Pascal)	1.013 250 E + 05
atmosphere (technical)	Pa	9.806 650 E + 04
bar	Pa	1.000 000 E + 05
barrel (petroleum, =42 gal)	m^3	1.589 873 E − 01
calorie (mean)	J	4.190 02 E + 00
centi Poise (cP)	Pa·s	1.000 000 E − 03
centi Stokes (cSt)	m^2/s	1.000 000 E − 06
degree Celsius[1]	K (Kelvin)	$T_K = T°C + 273.15$
degree Fahrenheit	K (Kelvin)	$T_K = (T°F + 459.67)/1.8$
degree Fahrenheit	°C	$T_{°C} = (T°F − 32)/1.8$
dyne	N	1.000 000 E − 05
foot (1 ft=1200/3937 m)	m	3.048000 E − 01
gravitate (g)	m/s^2	9.806 650 E + 00
gallon (UK)	m^3	4.546 090 E − 03
gallon (US)	m^3	3.785 412 E −03
horsepower (=550 ft·lbf/s)	W (watt)	7.456 999 E + 02
Inch	m	2.540 000 E − 02
kgf (=kilogram-force)	N	9.806 650 E + 00
lbf (=lbs, poundforce)	N	4.448 222 E + 00
pound (lb)	kg	4.535 924 E − 01

TABLE A1.4   (continued)
SI Conversion of US Units to SI Units

Convert from	To	Multiply
pound force (lbf, lbs)	N	4.448 222 E 00
mile (Int. = 5280 ft)	m	1.609 344 E + 03
mile (US nautical)	m	1.852 000 E + 03
mmHg (Torr)	Pa	0.133 322 E + 03
pint (US liquid)	m^3	4.731 765 E − 04
Poise	Pa · s	1.000 000 E − 01
psi (=pounds/square inch)	Pa	6.894 757 E +03
Rhe	1/(Pa · s)	1.000 000 E + 01
Slug	kg	1.459 390 E + 01
Stokes (St. kinematic visc.)	m^2/s	1.000 000 E − 04
torr (=mmHg, 0°C)	Pa	1.333 22 E +02
W (Watt)	J/s	1.000 000 E + 00
yard (yd)	m	9.144 000 E − 01

TABLE A1.5
Estimated Mechanical Properties of Nonresulurized Carbon Steel Bars

UNS No.	ANSI/SAE No.	Processing Method	Tensile Strength		Yield Strength		Elongation in 2 in., %	Reduction in Area, %	Brinell Hardness
			MPa	kpsi	MPa	kpsi			
G10060	1006	Hot rolled	300	43	170	24	30	55	86
		Cold drawn	330	48	280	41	20	45	95
G10100	1010	Hot rolled	320	47	180	26	28	50	95
		Cold drawn	370	53	300	44	20	40	105
G10160	1016	Hot rolled	380	55	210	30	25	50	111
		Cold drawn	420	61	350	51	18	40	121
G10180	1018	Hot rolled	400	58	220	32	25	50	116
		Cold drawn	440	64	370	54	15	40	126
G10200	1020	Hot rolled	380	55	210	30	25	50	111
		Cold drawn	420	61	350	51	15	40	121
G10300	1030	Hot rolled	470	68	260	37.5	20	42	137
		Cold drawn	520	76	440	64	12	35	149
G10350	1035	Hot rolled	500	72	270	39.5	18	40	143
		Cold drawn	550	80	460	67	12	35	163
G10400	1040	Hot rolled	520	76	290	42	18	40	149
		Cold drawn	590	85	490	71	12	35	170
G10450	1045	Hot rolled	570	82	310	45	16	40	163
		Cold drawn	630	91	530	77	12	35	179
10500	1050	Hot rolled	620	90	340	49.5	15	35	179
		Cold drawn	690	100	580	84	10	30	197
G10600	1060	Hot rolled	680	98	370	54	12	30	201
G10740	1074	Hot rolled	720	105	400	58	12	30	217
G10800	1080	Hot rolled	770	112	420	61.5	10	25	229
G10950	1095	Hot rolled	830	120	460	66	10	25	248

Source: Adapted from *ASM International Metals Handbook*, 10th ed. Contents: v. 1. Properties and Selection: Irons, Steels, and High-Performance Alloys; Engineer's Handbook: http://www.engineershandbook.com/Tables/materials.htm

TABLE A1.6

Minimum Room-Temperature Mechanical Properties of Austenitic Stainless Steels

ANSI (UNS)No.	Product Form	Processing Method	Tensile Strength MPa	Tensile Strength kpsi	Yield Strength MPa	Yield Strength kpsi	Elongation in 2 in, %	Reduction in Area, %	Brinell Hardness
201 (S20100)	P, Sh, St	Annealed	655	95	310	45	40		100 max.
	Sh, St	1/4 hard	860	125	515	75	25		
	Sh, St	1/2 hard	1030	150	760	110	18		
	Sh, St	3/4 hard	1210	175	930	135	12		
	Sh, St	Full hard	1280	185	965	140	9		
202 (S20200)	P, Sh, St	Annealed	620	90	260	38	40		
	Sh, St	1/4 hard	860	125	515	75	12		
301 (S30100)	B, P, Sh, St	Annealed	515	75	205	30	40		92 max.
	B, P, Sh, St	1/4 hard	860	125	515	75	25		
	B, P, Sh, St	1/2 hard	1030	150	760	110	18		
	B, P, Sh, St	3/4 hard	1210	175	930	135	12		
	B, P, Sh, St	Full hard	1280	185	965	140	9		
302 (S30200)	P, Sh, St	Annealed	515	75	205	30	40		92 max.
	B, P, Sh, St	1/4 hard	860	125	515	75	10		
	B, P, Sh, St	1/2 hard	1030	150	760	110	10		
	B, P, Sh, St	3/4 hard	1205	175	930	135	6		
	B, P, Sh, St	Full hard	1275	185	965	140	4		
304 (S30400)	W	Annealed	515	75	205	30	35	50	
304L (S30403)	F	Annealed	450	65	170	25	40	50	
316 (S31600)	P, Sh, St	Annealed	515	75	205	30	40		95 max.
316L (S31603)	P, Sh, St	Annealed	485	70	170	25	40		95 max.
416 (S41600)	F	Annealed	485	70	275	40	20	45	
430 (S43000)	P, Sh, St	Annealed	450	65	205	30	22		88 max.

Source: Adapted from *ASM International Metals Handbook*, 10th ed. Contents: v. 1. Properties and Selection: Irons, Steels, and High-Performance Alloys; Engineer's Handbook: http://www.engineershandbook.com/Tables/materials.htm

TABLE A1.7
Typical Mechanical Properties of Various Aluminum Alloys

Wrought No.	Temper	Tensile MPa	Strength kpsi	Yield MPa	Strength kpsi	Fatigue MPa	Strength kpsi	Elongation in 2 in., %	Brinell Hardness
1100	O	90	13	35	5	35	5	45	23
2011	T3	380	55	295	43	125	18	15	95
2017	T4	425	62	275	40	125	18	22	105
2024	T3	485	70	345	50	140	20	16	120
3003	H16	180	26	170	25	70	10	14	47
5052	H32	230	33	195	28	115	17	18	60
6061	T6	310	45	275	40	95	14	17	95
6063	T5	185	27	145	21	70	10	12	60
7075	T6	570	83	505	73	160	23	11	150

Source: Adapted from *ASM International Metals Handbook*, 10th ed. Contents: v. 2. Properties and Selection.— Nonferrous Alloys and Special-Purpose Materials; Engineer's Handbook: http://www.engineers handbook.com/Tables/materials.htm

TABLE A1.8
Composition for Typical Cast Irons

Type of Iron	Composition (%)				
	Carbon (C)	Silicon (Si)	Manganese (Mn)	Sulfur (S)	Phosphorus (P)
Gray (FG)	2.5–4.0	1.0–3.0	0.2–1.0	0.02–0.25	0.002–1.0
Ductile (SG)	3.0–4.0	1.8–2.8	0.1–1.0	0.01–0.03	0.01–0.1
Compacted graphite (CG)	2.5–4.0	1.0–3.0	0.2–1.0	0.01–0.03	0.01–0.1
Malleable (cast white) (TG)	2.2–2.9	0.9–1.9	0.15–1.2	0.02–0.2	0.02–0.2
White	1.8–3.6	0.5–1.9	0.25–0.8	0.06–0.2	0.06–0.2

Source: Reprinted with permission of ASM International. All rights reserved. www.asminternational.org

TABLE A1.9

Typical Mechanical Properties of As-Cast Standard Gray Iron Test Bars

ASTM A 48 class	Tensile Strength		Compressive Strength		Torsional Shear Strength		Modulus of Elasticity				Endurance Limit		Brinell Hardness
							Tensional		Torsional				
	MPa	kpsi	MPa	kpsi	MPa	kpsi	GPa	Mpsi	GPa	Mpsi	MPa	kpsi	
20	152	22	572	83	179	26	66–97	9.6–14.0	27–39	3.9–5.6	69	10	156
25	179	26	669	97	220	32	79–102	11.5–14.8	32–41	4.6–6.0	79	11.5	174
30	214	31	752	109	276	40	90–113	13.0–16.4	36–45	5.2–6.6	97	14	210
35	252	36.5	855	124	334	48.5	100–119	14.5–17.2	40–48	5.8–6.9	110	16	212
40	293	42.5	965	140	393	57	110–138	16.0–20.0	44–54	6.4–7.8	128	18.5	235
50	362	52.5	1130	164	503	73	130–157	18.8–22.8	50–55	7.2–8.0	148	21.5	262
60	431	62.5	1293	187.5	610	88.5	141–162	20.4–23.5	54–59	7.8–8.5	169	24.5	302

Source: Reprinted with permission of ASM International, All rights reserved. www.asminternational.org

---

TABLE A1.10
Coefficient of Static Friction

Material	Clean and Dry	Thick Oxide Film	Sulfide Film
Steel-Steel	0.78	0.27	0.39
Copper-Copper	1.21	0.76	0.74

TABLE A1.11
Centroids and Mass Moment of Inertia of Common Geometric Shapes

Shapes	Centroid	Moment of Inertia

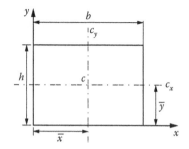

$$\bar{x} = \frac{b}{2}$$

$$\bar{y} = \frac{h}{2}$$

$$I_{c_x} = \frac{bh^3}{12}$$

$$I_x = \frac{bh^3}{12}$$

$$I_y = \frac{b^3 h}{12}$$

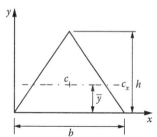

$$\bar{y} = \frac{h}{3}$$

$$I_{c_x} = \frac{bh^3}{36}$$

$$I_x = \frac{bh^3}{12}$$

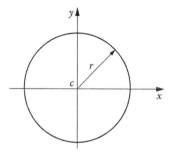

$$\bar{x} = 0$$

$$\bar{y} = 0$$

$$I_x = I_y = \frac{\pi r^4}{4}$$

$$J = \frac{\pi r^4}{2}$$

continued

TABLE A1.11   (continued)
Centroids and Mass Moment of Inertia of Common Geometric Shapes

Shapes	Centroid	Moment of Inertia
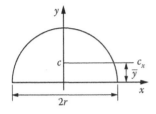	$\bar{x} = 0$ $\bar{y} = \dfrac{4r}{3\pi}$	$I_x = \dfrac{\pi r^4}{8}$ $I_{c_x} = 0.11 r^4$ $J_{c_y} = \dfrac{\pi r^4}{8}$
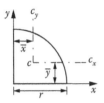	$\bar{x} = \dfrac{4r}{3\pi}$ $\bar{y} = \dfrac{4r}{3\pi}$	$I_x = I_y = \dfrac{\pi r^4}{16}$ $I_{c_x} = I_{c_y} = \dfrac{\pi r^4}{18.18}$

TABLE A1.12
Mass Moment of Inertia of Different Shapes

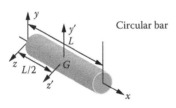

Circular bar

$$I_y = I_z = \frac{1}{3} m L^2$$

$$I_{y'} = I_{z'} = \frac{1}{12} m L^2$$

Circular cyclinder

$$I_z = I_{z'} = \frac{1}{2} m r^2$$

$$I_x = I_y = \frac{1}{12} m L^2 + \frac{1}{4} m r^2$$

$$I_{x'} = I_{y'} = \frac{1}{3} m L^2 + \frac{1}{4} m r^2$$

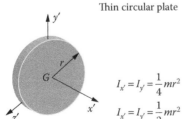

Thin circular plate

$$I_{x'} = I_{y'} = \frac{1}{4} m r^2$$

$$I_x = I_y = \frac{1}{2} m r^2$$

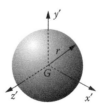

Sphere

$$I_{x'} = I_y = I_z = \frac{2}{5} m r^2$$

TABLE A1.12 (continued)
Mass Moment of Inertia of Different Shapes

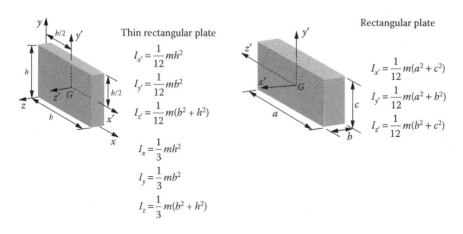

Thin rectangular plate

$$I_{x'} = \frac{1}{12}mh^2$$

$$I_{y'} = \frac{1}{12}mb^2$$

$$I_{z'} = \frac{1}{12}m(b^2 + h^2)$$

$$I_x = \frac{1}{3}mh^2$$

$$I_y = \frac{1}{3}mb^2$$

$$I_z = \frac{1}{3}m(b^2 + h^2)$$

Rectangular plate

$$I_{x'} = \frac{1}{12}m(a^2 + c^2)$$

$$I_{y'} = \frac{1}{12}m(a^2 + b^2)$$

$$I_{z'} = \frac{1}{12}m(b^2 + c^2)$$

TABLE A1.13
Deflections and Slopes of Simple Beams

Support System and Applied Load	Slope at End Point B	Maximum Deflection at B
(beam with load P at center, $x = L/2$)	$\theta_A = \theta_B = \dfrac{PL^2}{16EI}$	$y_{max} = \dfrac{PL^3}{48EI}$ (at the center where $x = L/2$)
(beam with load P at distance a, b)	$\theta_A = \dfrac{Pb(L^2 - b^2)}{6lEI}$ $\theta_B = \dfrac{Pa(L^2 - a^2)}{6lEI}$	$y_{center} = \dfrac{Pb(3L^2 - 4b^2)}{48EI}$ If $a \ge b$ $y_{center} = \dfrac{Pa(3L^2 - 4a^2)}{48EI}$ $y_{max} = \dfrac{Pb(L^2 - b^2)^{3/2}}{9\sqrt{3}LEI}$ (note that y_{max} is not at $x = L/2$)

continued

TABLE A1.13 (continued)
Deflections and Slopes of Simple Beams

Support System and Applied Load	Slope at End Point B	Maximum Deflection at B
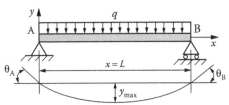	$\theta_A = \theta_B = \dfrac{qL^2}{24EI}$	$y_{max} = \dfrac{5qL^4}{384EI}$ (at the center where $x = L/2$)
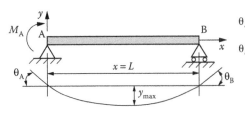	$\theta_A = \dfrac{ML}{3EI}$ $\theta_B = \dfrac{ML}{6EI}$	$y_{center} = \dfrac{ML^2}{16EI}$ $y_{max} = \dfrac{ML^2}{9\sqrt{3}EI}$ (at $x = L/\sqrt{3}$)

TABLE A1.14
Deflections and Slopes of Fix End Beams

Support System and Applied Load	Slope at End Point B	Maximum Deflection at B
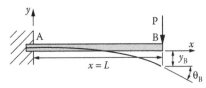	$\theta_B = \dfrac{PL^2}{3EI}$	$y_B = \dfrac{PL^3}{3EI}$
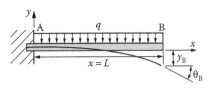	$\theta_B = \dfrac{qL^3}{6EI}$	$y_B = \dfrac{qL^4}{8EI}$

TABLE A1.14 (continued)
Deflections and Slopes of Fix End Beams

Support System and Applied Load	Slope at End Point B	Maximum Deflection at B

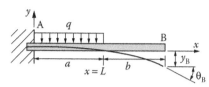

$$\theta_B = \frac{qa^3}{6EI}$$

$$y_B = \frac{qa^3}{24EI}(4L - a)$$

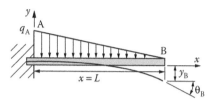

$$\theta_B = \frac{q_A L^3}{24EI}$$

$$y_B = \frac{q_A L^4}{30EI}$$

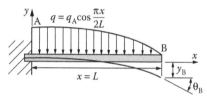

$$\theta_B = \frac{q_A L^3}{\pi^3 EI}(\pi^2 - 8)$$

$$y_B = \frac{2q_A L^4}{3\pi^4 EI}(\pi^3 - 24)$$

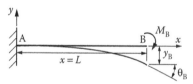

$$\theta_B = \frac{M_B L}{EI}$$

$$y_B = \frac{M_B L^2}{2EI}$$

$$\theta_B = \frac{M_B a}{EI}$$

$$y_B = \frac{M_B a}{2EI}(2L - a)$$

Appendix 2: Figures

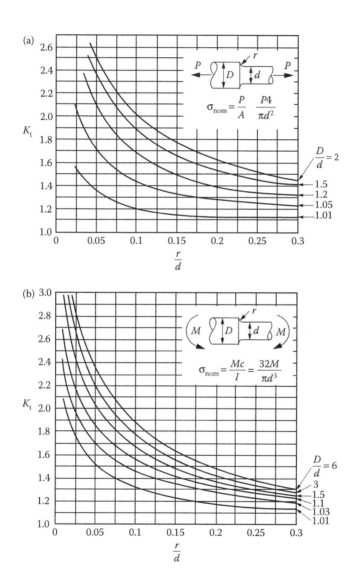

FIGURE A2.1 Stress concentration factors, K_t. (a) Shaft with fillet in axial load. (b) Shaft with fillet in bending. (From Peterson, R.E., *Stress Concentration Design Factors*, 1953. Copyright Wiley-VCH Verlag GmbH & Co. KGaA. Wiley, New York. Reproduced with permission.)

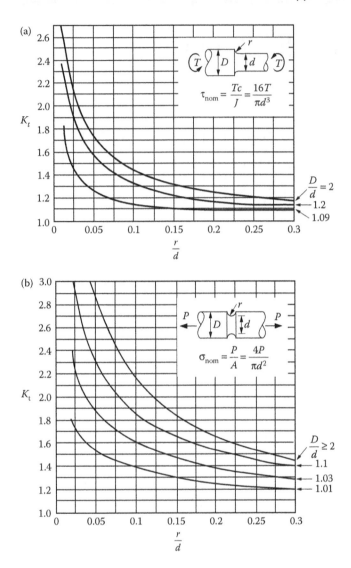

FIGURE A2.2 Stress concentration factors, K_t. (a) Shaft with fillet in Torsion. (b) Groved shaft in axial load. (From Peterson, R.E., *Stress Concentration Design Factors*, 1953. Copyright Wiley-VCH Verlag GmbH & Co. KGaA. Wiley, New York. Reproduced with permission.)

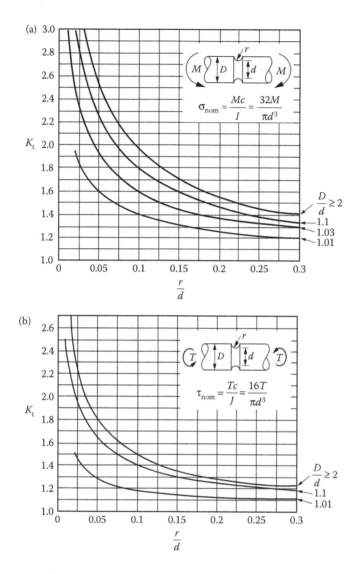

FIGURE A2.3 Stress concentration factors, K_t. (a) Groved shaft in bending. (b) Groved shaft in torsion. (From Peterson, R.E., *Stress Concentration Design Factors*, 1953. Copyright Wiley-VCH Verlag GmbH & Co. KGaA. Wiley, New York. Reproduced with permission.)

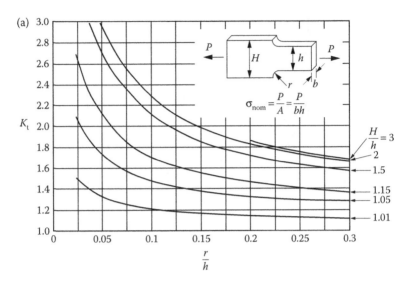

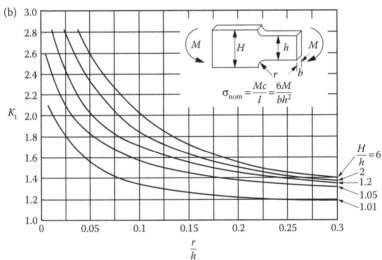

FIGURE A2.4 Stress concentration factors, K_t. (a) Flat plate with fillets in axial load. (b) Flat plate with fillets in bending. (From Peterson, R.E., *Stress Concentration Design Factors*, 1953. Copyright Wiley-VCH Verlag GmbH & Co. KGaA. Wiley, New York. Reproduced with permission.)

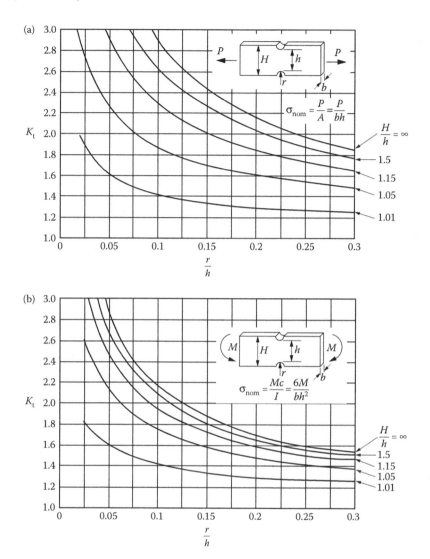

FIGURE A2.5 Stress concentration factors, K_t. (a) Notched flat plate in axial load. (b) Notched flat plate in bending. (From Peterson, R.E., *Stress Concentration Design Factors*, 1953. Copyright Wiley-VCH Verlag GmbH & Co. KGaA. Wiley, New York. Reproduced with permission.)

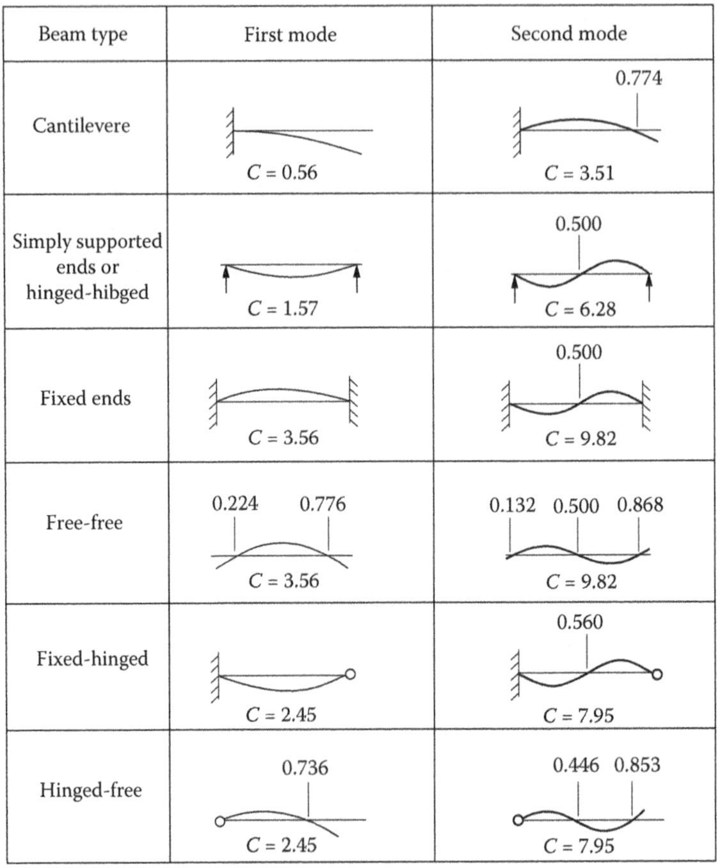

Beam type	First mode	Second mode
Cantilevere	$C = 0.56$	0.774 $C = 3.51$
Simply supported ends or hinged-hibged	$C = 1.57$	0.500 $C = 6.28$
Fixed ends	$C = 3.56$	0.500 $C = 9.82$
Free-free	0.224 0.776 $C = 3.56$	0.132 0.500 0.868 $C = 9.82$
Fixed-hinged	$C = 2.45$	0.560 $C = 7.95$
Hinged-free	0.736 $C = 2.45$	0.446 0.853 $C = 7.95$

$f_n = C\sqrt{\dfrac{EIg}{wL^4}}$ where C is the modal constant and $w \rightarrow lb/in$

FIGURE A2.6   Resonant frequencies of uniform beams.

Support type	Natural frequency equations
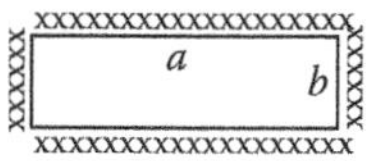	$$f_n = \frac{\pi}{2}\left(\frac{D}{\rho}\right)^{1/2}\left(\frac{1}{a^2}+\frac{1}{b^2}\right)$$
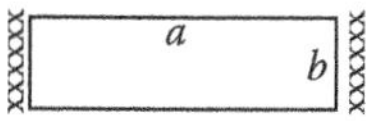	$$f_n = \frac{\pi}{2a^2}\left(\frac{D}{\rho}\right)^{1/2}$$
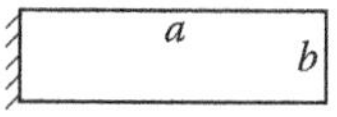	$$f_n = \frac{0.56}{a^2}\left(\frac{D}{\rho}\right)^{1/2}$$
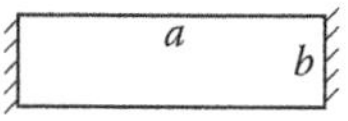	$$f_n = \frac{3.55}{a^2}\left(\frac{D}{\rho}\right)^{1/2}$$
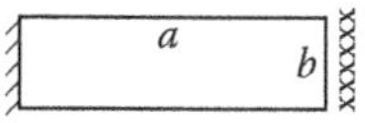	$$f_n = \frac{0.78\times\pi}{a^2}\left(\frac{D}{\rho}\right)^{1/2}$$
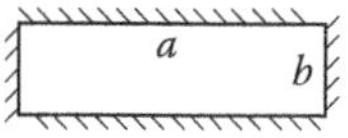	$$f_n = \frac{\pi}{1.5}\left[\frac{D}{\rho}\left(\frac{3}{a^+}+\frac{2}{a^2+b^2}+\frac{3}{b^4}\right)\right]^{1/2}$$

——— → Free edge
ᴧᴧᴧᴧ → Fixed edge
xxxxxx → Simply supported edge

FIGURE A2.7   Natural frequency equations for uniform plates.

REFERENCE

1. Peterson, R.E., *Stress Concentration Design Factors*, Wiley, New York, 1953.

Index

Printed and bound by CPI Group (UK) Ltd, Croydon, CR0 4YY

21/10/2024

01777107-0009